Hyphenated Techniques in Speciation Analysis

RSC Chromatography Mongraphs

Series Editor: Roger M. Smith, *University of Technology, Loughborough, UK*
Advisory Panel: J.C. Berridge, *Sandwich, UK*; G.B. Cox, *Illkirch, France*; I.S. Lurie, *Virginia, USA*; P.J. Schoenmaker, *Amsterdam, The Netherlands*; C.F. Simpson, *London, UK*; G.G. Wallace, *Wollongong, Australia*.

This series is designed for the individual practising chromatographer, providing guidance and advice on a wide range of chromatographic techniques with the emphasis on important practical aspects of the subject.

Supercritical Fluid Chromatography
edited by Roger M. Smith, *University of Technology, Loughorough, UK*

Packed column SFC
by T.A. Berger, *Berger Instruments, Newark, Delaware, USA*

Chromatographic Integration Methods, Second Edition
by Norman Dyson, *Dyson Instruments Ltd, UK*

Separation of Fullerenes by Liquid Chromatography
edited by K. Jinno, *Toyohashi University of Technology, Japan*

HPLC: A Practical Guide
by Toshihiko Hanai, *Health Research Foundation, Kyoto, Japan*

Applications of Solid Phase Microextraction
Edited by Janusz Pawliszyn, *University of Waterloo, Ontario, Canada*

Capillary Electrochromatography
edited by Keith D. Bartle, *University of Leeds, UK*
and Peter Myers, *X-tec Consulting Ltd, UK*

Cyclodextrins in Chromatography
By T. Cserhati and E. Forgacs, *Chemical Research Centre, Hungarian Academy of Sciences, Budapest, Hungary*

How to obtain future titles on publication

A standing order plan is available for this series. A standing order will bring delivery of each new volume upon publication. For further information please contact:
Sales and Customer Care, Royal Society of Chemistry, Thomas Graham House, Science Park, Milton Road, Cambridge CB4 0WF
Telephone: +44(0) 1223 432360 E-mail: sales@rsc.org

RSC
CHROMATOGRAPHY
MONOGRAPHS

Hyphenated Techniques in Speciation Analysis

Joanna Szpunar,
Centre National de la Recherche Scientifique (CNRS), Pau, France

Ryszard Łobiński,
*Centre National de la Recherche Scientifique (CNRS), Pau,
France, and Warsaw University of Technology, Warsaw, Poland*

advancing the chemical sciences

ISBN 0-85404-545-7

A catalogue record for this book is available from the British Library

Published by The Royal Society of Chemistry,
Thomas Graham House, Science Park, Milton Road,
Cambridge CB4 0WF, UK

Registered Charity Number 207890

For further information see our web site at www.rsc.org

Typeset by Keytec Typesetting Ltd, Dorset, UK
Printed by Athenaeum Press Ltd, Gateshead, Tyne and Wear, UK

To our children: Izabella, Gabrysia and Robert

Preface

The last two decades have brought a surge of interest in speciation analysis, a field of trace element analytical chemistry that deals with the detection, identification and determination of individual chemical forms of metals and metalloids. Indeed, it is now generally accepted by environmental chemists, nutritionists and toxicologists that information on the total element concentration in the sample is not only insufficient to evaluate its toxicity, essentiality or bioavailability, but may even be misleading. The increasing awareness of the importance of elemental speciation is resulting in a growing demand from research and routine laboratories for analytical techniques capable of providing species-specific information for the environment, agriculture and nutrition, clinical chemistry and toxicology, medicine and pharmacology and industrial process chemistry.

Hyphenated techniques, based on the combination of high resolution separation techniques with element or molecule specific detectors, represent a unique analytical tool able to provide qualitative and quantitative information on element species at trace and ultratrace levels in complex matrices. A growing number of chromatographic and electrophoretic separations can be efficiently coupled with element specific detection, *e.g.* atomic emission or inductively coupled mass spectrometry as well as with molecule specific detection, *e.g.* electrospray mass spectrometry. Couplings such as GC-AAS, GC-MIP AES and GC-ICP MS, HPLC-ICP MS, CZE-ICP MS and HPLC or CZE-electrospray MS/MS, have become well established tools for elemental speciation analysis.

The field of speciation analysis itself has been undergoing a continuous evolution. The classical activities have involved species-specific determination of anthropogenic organometallic contaminants: organolead, organomercury or organotin compounds, and products of their environmental degradation. These are giving way to a search for endogenous metal and metalloid species, that are present in living organisms as a consequence of the biochemical evolution or have been bio-induced in response to a metal stress. In terms of analytical developments the demonstration of the analytical craft and skills of an analyst to determine a particular elemental species in a sample is being replaced by exploratory investigations aimed at the detection of unknown elemental species in the tissues of a living organisms, their identification and/or structural characterisation.

The book is intended as not only an introductory text to newcomers to the field of elemental speciation analysis. It also offers a critical overview of the research carried out in the field that may serve an already practising analyst. The book is organised in two parts: the first is focused on the technical aspects of the different analytical techniques available and the second on their application to analytical problems in different disciplines. On the subject of analytical techniques individual chapters are devoted to gas chromatography, liquid chromatography and electrophoretic techniques with element specific detection, and electrospray mass spectrometry. The importance of quality control and assurance in speciation analysis is reflected by a dedicated chapter. On applications the selection and organisation of chapters reflect the different maturity of various research areas. The methodology for speciation analysis of methylated species, organolead, -tin and -mercury is well established so these chapters are focused on validated methods that are being implemented in routine laboratories. Regarding naturally occurring metallospecies preference has been given to approaches that have been allowing the exploration of the field in terms of the detection, characterisation and identification of new metallobiomolecules.

The representative coverage of the many facets of this broad and dynamically evolving field has been a difficult task. Indeed, during recent years speciation has become a fashionable area of inorganic trace element research that has resulted in the exponential proliferation of research and review publications. Over 2000 speciation-related papers have been published, many of which have unfortunately only contributed to the information noise. Therefore, the techniques, methods and applications discussed in this book had to be and are a critical selection from the massive literature available. The choice was made on the basis of our practical experience gathered over the last 12 years of research in the field where we have had the opportunity to follow the development of analytical methods in real time and to develop, test or adopt many of the applications in our laboratory.

Joanna Szpunar and Ryszard Łobiński
Pau, December 2002

Contents

Terms and Abbreviations

AAS	atomic absorption spectrometry
AE	anion exchange
AED	atomic emission detection
AES	atomic emission spectrometry
AFS	atomic fluorescence spectrometry
CE	cation exchange
CEC	capillary electrochromatography
CID	collision-induced dissociation
CRM	certified reference material
CT	cryogenic trap
CZE	capillary zone electrophoresis
DBT	dibutyltin
DDT	dithiothreitol
DDTC	diethyl dithiocarbonate
DFO	desferioxamine
DIN	direct injection nebuliser
DL	detection limit
DMAA	dimethylarsonic acid
DPhT	diphenyltin
ECD	electron capture detection
EI	electron impact
ES	electrospray
ET	electrothermal
ETV	electrothermal vaporisation
FAAS	flame atomic absorption spectrometry
FAB	fast atom bombardment
FID	flame ionisation detection
FPD	flame photometric detection
GC	gas chromatography
GE	gel electrophoresis
GF	graphite furnace

HG	hydride generation
HHPN	hydraulic high pressure nebulisation
HPLC	high performance liquid chromatography
HR	high resolution
ICP	inductively coupled plasma
ID	isotope dilution
IEF	isoelectric focusing
INAA	instrumental neutron activation analysis
LA	laser ablation
LC	liquid chromatography
LT	low temperature
MALDI	matrix assisted laser desorption ionization
MBT	monobutyltin
MMAA	monomethylarsinic acid
MPhT	monophenyltin
MC GC	multicapillary gas chromatography
MIP	microwave induced plasma
MS	mass spectrometry
MS/MS	tandem mass spectrometry
MT	metallothionein
NMR	nuclear magnetic resonance
PAGE	polyacrylamide gel electrophoresis
PC	phytochelatin
PFPD	pulsed flame photometric detection
PIXE	proton induced X-ray emission
QF	quartz furnace
Q MS	quadrupole mass spectrometer
SBSE	stir bar sorptive extraction
SCID	source collision induced dissociation
SDS	sodium dodecylsulphonate
SE	size exclusion
SEC	size exclusion chromatography
SFC	supercritical fluid chromatography
SPME	solid phase micro extraction
TBT	tributyltin
TMAH	tetramethylammonium hydroxide
TPrT	tripropyltin
TPhT	triphenyltin
TOF	time of flight
TLC	thin layer chromatography
TXRF	total-reflection X-ray fluorescence
UV	ultraviolet
XRF	X-ray fluorescence

Part I Principles and Fundamentals

CHAPTER 1

The Concept of Speciation Analysis and Hyphenated Techniques

1 Introduction

The analysis of metal(loid) organic compounds has become increasingly important in the last decade because the organic species of some elements have turned out to be much more toxic than their inorganic forms. The origin of metal species can be either anthropogenic input (*e.g.* as pesticides) or a result of biological transformations of inorganic forms of elements by living organisms. The harmful effects of trace metal species were fully recognised after the death of 50 residents of the Minamata fishing village in Japan who had experienced the biomethylation of mercury in their everyday food.[1] The spillage of tetraalkyllead in the Mediteranean due to the M/S Cavteat accident made the analytical community sensitive to organic forms of lead.[2] Extinction of the oyster population in the Arcachon Bay in Southern France stimulated interest in the possible release of butyltins from antifouling paints.[3] The above cases raised awareness of the importance of knowing the concentration of a particular species, defined the suspected analyte and analytical sample and, consequently, stimulated the development of analytical methodology.

It has also been widely recognised in biochemistry that the proper functioning of life is critically dependent on trace elements in a number of different ways. Some metals (*e.g.* Hg, Pb) and metalloids (As) are highly toxic whereas others (*e.g.* Mo, Mn, Fe, Co, Cu, Zn), considered essential, are needed for the accomplishment of life processes.[4] A number of other elements (*e.g.* V, Cr, Ni) are recognised as being beneficial to life. From a chemical point of view the intake, accumulation, transport and storage of essential or toxic metals and metalloids are realised by surrounding the element ion by electron pair donating biological ligands. Sometimes this process is accompanied by the synthesis of specific ligands such as metallothioneins, or by the formation of a metalloid-carbon bond as in the case of selenoamino acids or organoarsenic compounds. Since the evolution of a metal in a living organism happens by its interaction with the highly complex coordinating environment and involves a number of species with different properties, information on the total elemental concentration in a biotissue is not sufficient and may even be misleading in understanding the metabolism, bioavailability and toxicity of a metal or a metalloid.

Classical analytical approaches based on high performance liquid chromatography (HPLC) with ultraviolet (UV) detection, and gas chromatography (GC) with flame ionisation (FID) or electron capture (ECD) detection have become insufficient to efficiently address the increasing challenges of metal species analysis in terms of species-specificity and sensitivity. The origins of modern speciation analyses are closely related to a wider use of element or molecule specific detection in chromatography and electrophoresis, the developments referred to as hyphenated techniques.

2 Speciation Analysis: The Definition

The term speciation was first introduced by biologists to describe the evolution of species. Referring to this evolutionary concept, geochemists and environmental chemists have often applied the word speciation to the transformations taking place during cycling of the elements. An example is the changes that occur between the leaching of trace elements from soil or rock and their subsequent distribution in the aquatic environment. The term speciation has been used since the 50s by aquatic geochemists to distinguish between 'dissolved metal' and 'particulate metal' in order to improve the understanding of the metal transport in waterways. A simple filtration through a 0.45 μm membrane allowed discrimination between the two different phases. The simultaneous and rapid development of electrochemistry enabled the analyst to distinguish between free and complexed metal species in the dissolved fraction.

This philosophy was developed by two well known works often cited in the context of environmental speciation analysis. One is Tessier's sequential extraction scheme developed for the determination of the metal partitioning between the various mineral phases of environmental particulate material.[5] The other is Florence's scheme implying the differentiation between different metal fractions in water using preconcentration with Chelex-100 ion-exchange resin and final detection by electrochemistry or atomic absorption spectrometry (AAS).[6]

The common feature of these works was the impossibility of distinguishing between individual metal-containing species at the molecular level. They dealt with metal fractions and not with metal species. Truly specific information was first acquired by Kölb who coupled chromatography for the separation of the species of interest with atomic spectrometry for the sensitive and selective detection of the metal.[7] This concept was popularized by van Loon in an A-page Analytical Chemistry article in 1979 which initiated an exponential growth in the number of papers on speciation analysis.[8]

In an attempt to end the confusion regarding the usage of the term *speciation*, an IUPAC Interdivisional Working Party recommended the use of this term to describe the distribution of an element amongst defined chemical species in a system.[9] A species was defined as a form of an element specified as to isotopic composition, electronic or oxidation state, and/or complex or molecular structure. According to the same definition speciation analysis denoted the analytical activities of identifying and/or measuring the quantities of one or more individual chemical species in a sample.[9]

The term speciation is different from fractionation which describes the process of classification of an analyte or a group of analytes from a certain sample according to physical (*e.g.* size, solubility) or chemical (*e.g.* bonding, reactivity) properties.

3 Occurrence and Classification of Metal Species

The variety of chemical species of interest in different disciplines is shown schematically in Figure 1.1. Basically they can be divided into exogeneous species, such as environmental contaminants or metallodrugs and products of their degradation, and endogenous species, such as natural metabolites of arsenic or metal complexes with bioligands. From a chemical point of view elemental species can be divided into redox states, organometallic species (containing a covalent carbon-metal bond) and coordination complexes. The latter include

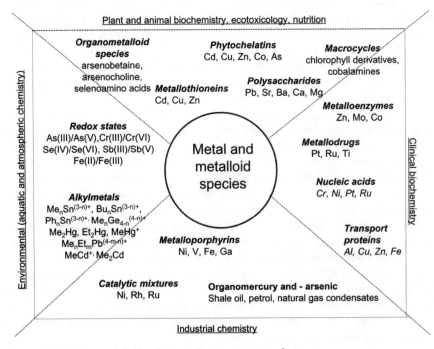

Figure 1.1 *Species and fields of interest in speciation analysis*

simple (*e.g.* halide) or complex (*e.g.* citrate, tartrate, oxalate, phytate, amino acids, oligopeptides) ligands, macrocyclic chelating ligands (*e.g.* porphyrins), or macromolecules (*e.g.* polypeptides, proteins, DNA restriction fragments, polysaccharides).

The principal classes of metal species of interest include:

- *Volatile organometallics in air and landfill gases* Volatile metal(loid) compounds have been identified in a variety of anthropogenic gases, including landfill gas and sewage sludge digester gas. These compounds are non-charged hydrides and/or methylated or alkylated compounds of main-group elements of groups 12 to 17. Compounds such as dimethylmercury (Me_2Hg), dimethyl selenide (Me_2Se), methyl- and butyltin, trimethylstibine (Me_3Sb), trimethylbismuthine (Me_3Bi), methylated arsines (Me_xAsHy, $x + y = 3$), dimethyltelluride (Me_2Te), alkylated lead (Et_xMe_yPb, $x + y = 4$) have been identified in concentrations ranging from $ng\,m^{-3}$ to $\mu g\,m^{-3}$. Carbonyls of Ni, Mo and W have been found in sewage gas and landfill gas.[10] Natural environments, such as hot springs rich in algae can also produce volatile metalloid compounds.[11]

- *Anthropogenic organometallic contaminants* Organotin compounds, *e.g.* butyl- and phenyltins, are used in antifouling paints whereas tetraalkyllead compounds are still added in some countries as antiknock additives to petrol. The latter are being replaced by organomanganese additives, *e.g.* methylcyclopentadienylmanganese tricarbonyl. These tin and lead species, usually toxic, find their way into the environment raising the need for monitoring of both native species and products of their degradation.

- *Natural organometallics in shale oils, natural gas and condensate* Porphyrins, ubiquitously present in natural energy resources, such as shale oil or coal, have the facility of binding many metals, *e.g.* V, Ni, Fe, Ga, Ti in thermally stable complexes. Their fate becomes of concern during burning and processing of these materials. Speciation of mercury and arsenic, which undergo alkylation reactions in natural gas, is of concern during processing of gas and gas condensates.

- *Biosynthesized molecules with 'true' metal(metalloid) - carbon bonds* This category includes selenoamino acids and their higher analogues: selenopeptides and selenoproteins. They can further coordinate metals with *S*-affinity using the Se atom as the coordination center.[12] Another important class includes organoarsenic compounds: methylarsonic acids, quaternary compounds (*e.g.* arsenobetaine) and arsinoylriboside derivatives (arsenosugars).[13,14]

- *Complexes with biosynthesized macrocyclic chelating agents* The most important group is the analogues of tetrapyrrole which in their deprotonated forms can tightly bind even relatively labile divalent metal cations. The best known compounds of this group include chlorophyll and products of its degradation, cobalamins (the coenzymatically active forms of vitamin B12),[15] and porphyrins[16] including the heme group found in hemoglobin, myoglobin, cytochromes and peroxidases.

- *Complexes with nucleobases, oligo- and polynucleotides, and -nucleosides*
  Heterocyclic nucleobases, alone or as constituents of nucleosides or
  nucleotides, offer several different coordination sites for metal ions. Of
  particular interest is the coordination of metal ions *e.g.* CrO_4^- or inert
  metal complexes to DNA because of their specificity with regard to certain
  base-pair sequences in the double helix.

- *Complexes with amino acids, oligopeptides and polypeptides (proteins)*
  Metal complexes with proteins, including enzymes, are carriers of biochemical function. Whereas the carboxamide function of peptide bonds
  -C(=O)-N(-H)- is only a poor metal coordination site, peptides contain
  several functional groups in the side chains that are particularly well
  suited for metal coordination. They include especially cysteine ($-CH_2SH$)
  and methionine $-CH_2CH_2SCH_3$, which bind metals with sulfur affinity
  (Cd, Cu, Zn), and histidine of which both nitrogen atoms become
  available for coordination after metal-induced deprotonation (*e.g.* Cu, Zn
  in superoxide dismutase). Peptide-complexed metal ions are known to
  perform a wide variety of specific functions (regulatory, storage,
  catalytic, transport) associated with life processes. The greatest interest
  has been attracted by essential elements associated with ferritin (Fe, Cu,
  Zn), β-amylase (Cu), alcohol dehydrogenase (Zn), carbonic anhydrase
  (Cu, Zn) and other proteins. Homeostatic control, metabolism and
  detoxification of toxic elements (*e.g.* Cd, Hg) by their interaction with
  metallothioneins (MTs) have been in the focus of ecotoxicology and
  clinical chemistry.[17,18] Detoxification mechanisms of plants exposed to
  heavy metals involve the synthesis of small thiol peptides (phytochela-
  tines) able to chelate heavy metals due to the high cysteine content in
  the molecule.[19]

- *Complexes with other biomacromolecules (polysaccharides, glycoproteins)*
  Relatively little is known about the relevance of metal coordination to
  lipids and carbohydrates, although the potentially negatively charged
  oxygen functions can bind cations electrostatically and even undergo
  chelate coordination *via* polyhydroxy groups.[20] The complexation of
  divalent cations with the carboxylic acid groups of uronic acids from plant
  cell wall polysaccharides (pectins) is well established.[20]

- *Exogeneous species: metallodrugs* Platinum (cisplatin, carboplatin), Ru^{3+}
  (*fac*-$RuCl_3(NH_3)_3$) and gold (auranofin) compounds are well-known in
  cancer therapy whereas some other gold compounds (aurithiomalate,
  aurothioglucose) are important antiarthritic drugs.[21] A wide range of Tc
  compounds (*e.g.* Tc-labelled antibodies, Tc-mercaptoacetyl glycine complex) are used for diagnostic imaging of renal, cardiac and cerebral
  functions and of various forms of cancer.[22] Gadolinium (III) polyaminopolycarboxylic crown complexes are widely used as magnetic resonance
  imaging contrast reagents. Some vanadium compounds are antidiabetic
  agents. The analytical challenges include both the identification of products
  of metallodrug metabolism and the understanding of the binding of
  metallodrugs to transport proteins and DNA fragments.

4 The Concept of Hyphenated Techniques

Interest in species-selective instrumental analysis was raised amongst the analytical community by gas chromatographers who first appreciated the advantages of an element selective detector in the mid 60s.[7] As shown in Figure 1.2,[23] it allows the elimination of the non-specific background and thus a considerable

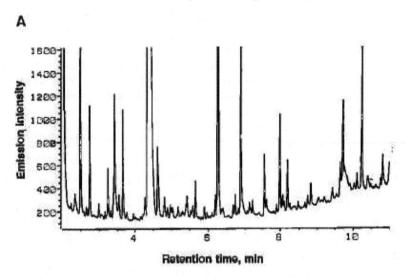

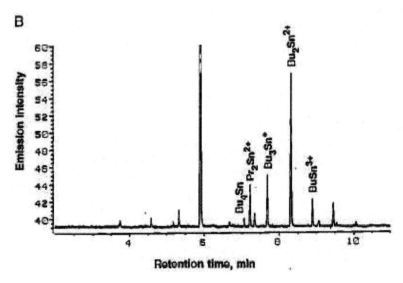

Figure 1.2 *Selectivity of an element specific detector in the gas chromatography of elemental species: chromatograms of a sediment sample extract (a) on the C-193 nm channel,(b) on the Sn-303 nm channel by GC-MIP AED*
(Reprinted with permission from *Anal. Chem.*, 1992, **64**, 159, copyright 1992, American Chemical Society)

increase in the signal-to-background ratio. Peaks in the chromatograms are due to metal-containing species only and can be identified on the basis of the retention time. The detection limits can be decreased to femtogram levels in the case of gas chromatography, and subpicogram levels in liquid chromatography. The sensitivity offers a considerable advantage over classical biochemical separation/purification protocols prior to structural analysis by X-ray diffraction, nuclear magnetic resonance (NMR), Mössbauer spectroscopy, of electronic, vibrational or circular dichroism spectroscopy which require fairly large amounts of well purified analyte compounds, typically in the milligram range.[24-26]

The term hyphenated techniques, introduced by Hirschfeld,[27] refers to an on-line combination of a chromatographic (later also electrophoretic) separation technique with a sensitive and element-specific detector (usually an atomic absorption, emission or mass spectrometer). The approach gained the particular attention in the speciation analysis of environmental organometallic pollutants (organotin, organolead, organomercury) and redox states, and has been the subject of a number of status papers,[28-40] edited works,[41-44] and fundamental and comprehensive reviews. The lattest were directed to the separation component of the hyphenated techniques (gas chromatography (GC),[45] supercritical fluid (SFC)[46] or liquid chromatography (LC)[47]), the detection component (inductively coupled plasma (ICP) atomic emission spectroscopy (AES) and ICP mass spectrometry (MS)),[33,48-50] or a particular coupling such as GC-ICP MS,[51] HPLC-ICP MS[52,53] or capillary zone eletrophoresis (CZE)-ICP MS.[54]

The various possibilities for the on-line coupling of a separation technique with an element (moiety, species) specific detector for bioinorganic speciation analysis include different types of HPLC or electrophoresis for separation, and atomic spectrometry (or molecular MS) for detection. The hyphenated techniques available for species-selective analysis are schematically shown in Figure 1.3. The

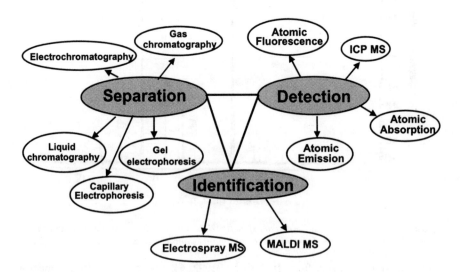

Figure 1.3 *Hyphenated techniques for speciation analysis*

presence of a metal incorporated or bound to an organic molecule in a sample is considered to be the prerequisite for using an element-specific detector. Nevertheless, some reports have indicated the possibility of employing a coupled technique for the analysis of a metal-free compound, provided that the latter is derivatised on-column or post-column by saturating the metal binding sites with a metal.[55,56]

5 The Choice of a Hyphenated Technique

The choice of hyphenated technique depends primarily on the research objective. The separation component of the coupled system becomes of particular importance when the target species have similar physicochemical properties. It may

Table 1 *Selection of hyphenated technique as a function of the analytical task*

Matrix	Analyte	Preferred separation technique	Preferred detection technique
Environment	methylmercury	GC	AFS, AAS, ICP MS
	organometallic contaminants	GC	MIP AES, ICP MS
	redox states	anion exchange HPLC, CZE	ICP MS
	humic and fulvic acids	size exclusion HPLC	ICP MS
Petrochemical (shale oil petrol, gas condensates)	alkylmercury, organoarsenic, metalloporphyrins	GC	ICP MS
Foodstuffs and food supplements	organometallic contaminant residues	GC	MIP AES, ICP MS
	selenium, selenomethionine	reversed phase HPLC, anion exchange HPLC	ICP MS
	arsenic	anion-exchange HPLC	ICP MS
Plant and animal biochemistry	Metal complexes with amino acids	CZE	ICP MS
	phytochelatins	size exclusion HPLC, reversed phase HPLC	ICP MS
	metallothioneins	size exclusion HPLC, reversed phase HPLC, CZE	ICP MS
	proteins	size exlusion HPLC	ICP MS
Cinical chemistry	metal complexes with proteins	size exclusion LC	ICP MS
	cobalamins, porphyrins	reversed phase HPLC	ICP MS
	metallodrugs	reversed phase HPLC, cation exchange HPLC	ICP MS

even be necessary to combine two or more separation mechanisms in series to assure that a unique species arrives at the detector at a given time. The choice of detector component becomes crucial when the concentration of analyte species in the sample is very small and low limits of detection are required. An important problem is often the interface between chromatography and spectrometry as the separation conditions may be not compatible in terms of flow rate and mobile phase composition with those required by the detector.

Usually, chromatography and spectrometry can be coupled on-line. However, when a polyacrylamide gel electrophoresis (PAGE) technique is used, off-line detection of metal species carried out directly in the gel[57–60] or after extraction (blotting) of proteins from the gel[61] is necessary. Also, the preference for a highly sensitive discrete atomisation technique such as electrothermal atomic absorption spectroscopy (ETAAS)[37,62–70] or electrothermal vapourisation (ETV) ICP MS[63] may be the reason for choosing an off-line method of coupling.

Table 1 summarizes the selection of a hyphenated technique as a function of the application objective.

In contrast to classical speciation analysis[71] where standards for most of the anthropogenic pollutants are available, the majority of species of interest in bioinorganic trace element speciation analysis have not yet been isolated in a sufficiently pure state to be used as retention or migration time standards. Therefore, it is becoming of paramount importance to employ in parallel a molecule (or moiety) specific detector to determine the identity of the eluted species. Mass spectrometry: fast atom bombardment (FAB MS),[72] electrospray (ES MS)[73–76] or matrix assisted laser desorption ionisation time of flight (MALDI TOF MS)[77] have been the viable choices.

References

1. M. Harada, *Crit. Rev. Toxicol.*, 1995, **25**, 1.
2. G.F. Harrison, *Lead in the Marine Environment*, in M. Branica and Z. Konrad (Eds.), Pergamon, Oxford, 1980.
3. C. Alzieu, J. Sanjuan, P. Michel, M. Borel and J.P. Dreno, *Mar. Pollut. Bull.*, 1989, **20**, 22.
4. D.M. Taylor and D.R. Williams, *Trace Element Medicine and Chelation Therapy*, Royal Society of Chemistry, Cambridge, 1995.
5. A. Tessier, P.G.C. Campbell and M. Bisson, *Anal. Chem.*, 1979, **7**, 844.
6. T.M. Florence, *Trends Anal.Chem.*, 1983, **2**, 162.
7. B. Kölb, G. Kemmner, F.H. Schleser and E. Wiedeking, *Fresenius' J. Anal. Chem.*, 1966, **221**, 166.
8. J.C. Van Loon, *Anal. Chem.*, 1979, **51**, 1139A.
9. D. Templeton, F. Ariese, R. Cornelis, L.G. Danielsson, H. Muntau, H.P. van Leeuven and R. Lobinski, *Pure Appl. Chem.*, 2000, **72**, 1453.
10. J. Feldmann, *J. Environ. Monit.*, 1999, **1**, 33.
11. A.V. Hirner, E. Krupp, F. Schulz, M. Koziol and W. Hofmeister, *J. Geochem. Explor.*, 1998, **64**, 133.

12. C. Sasakura and K.T. Suzuki., *J. Inorg. Biochem.*, 1998, **71**, 159.
13. Y. Shibata, M. Morita and K. Fuwa, *Adv. Biophys.*, 1992, **28**, 31.
14. M. Morita and J.S. Edmonds, *Pure Appl. Chem.*, 1992, **64**, 575.
15. H. Chassaigne and R. Lobinski, *Anal. Chim. Acta.*, 1998, **359**, 227.
16. J.W. Ho, *J. Liq. Chromatogr.*, 1990, **13**, 3741.
17. M.J. Stillman, C.F. Shaw and K.T. Suzuki, *Metallothioneins Synthesis, Structure and Properties of Metallothioneins, Phytochelatins and Metalthiolate Complexes*, Wiley, New York, 1992.
18. M.J. Stillman, *Coord. Chem. Rev.*, 1995, **144**, 461 .
19. M.H. Zenk, *Gene*, 1996, **179**, 21.
20. D.M. Whitfield, S. Stoijkovski and B. Sarkar, *Coord. Chem. Rev.*, 1993, **122**, 171.
21. B.K. Keppler, *Metal Complexes in Cancer Chemotherapy*, VCH, Weinheim, 1993.
22. O.K. Hjelstuen, *Analyst*, 1995, **120**, 863.
23. R. Lobinski, W. Dirkx, M. Ceulemans and F.C. Adams, *Anal. Chem.*, 1992, **64**, 159.
24. W. Kaim and B. Schwederski, *Bioinorganic Chemistry: Inorganic Elements in the Chemistry of Life*, Wiley, Chichester, 1994.
25. R.J.P. Williams, *Coord. Chem. Rev.*, 1990, **100**, 573.
26. S.J. Lippard and J. M. Berg, *Principles of Bioinorganic Chemistry*, University Science Books, Mill Valley, CA, 1994.
27. T. Hirschfeld, *Anal. Chem.*, 1980, **52**, 297A.
28. N.P. Vela and J.A. Caruso, *J. Anal. At. Spectrom.*, 1993, **8**, 787.
29. W. Lund, *Fresenius' J. Anal. Chem.*, 1990, **337**, 557.
30. N.P. Vela, L.K. Olson and J.A. Caruso, *Anal. Chem.*, 1993, **65**, 585A.
31. A. Seubert, *Fresenius' J. Anal. Chem.*, 1994, **350**, 210.
32. M.J. Tomlinson, L. Lin and J.A. Caruso, *Analyst*, 1995, **120**, 583.
33. P.C. Uden, *J. Chromatogr, A.*, 1995, **703**, 393.
34. S. Caroli, *Microchem J.*, 1995, **51**, 64.
35. S.G. Dai and C.R. Jia, *Anal. Sci.*, 1996, **12**, 355.
36. I. Havezov, *Fresenius' J. Anal. Chem.*, 1996, **355**, 452.
37. A.K. Das and R. Chakraborty, *Fresenius' J. Anal. Chem.*, 1997, **357**, 1.
38. B. Welz, *J. Anal. At. Spectrom.*, 1998, **13**, 413.
39. S.L. Bonchin Cleland, H. Ding and J.A. Caruso, *Am. Lab.*, 1995, **27**, 34N.
40. J. Szpunar and R. Lobinski, *Fresenius' J. Anal. Chem.*, 1999, **363**, 550.
41. I.S. Krull (Ed.), *Trace Metal Analysis and Speciation*, Elsevier, Amsterdam, 1991.
42. P.C. Uden (Ed.), *Element-Specific Chromatographic Detection by Atomic-Emission Spectroscopy*, ACS, Washington, D.C., 1991.
43. J.A. Caruso, K.L. Sutton and K.L. Ackley (Eds.), *Elemental Speciation. New approaches for trace element analysis*, Elsevier, Amsterdam, 2000.
44. L. Ebdon, L. Pitts, R. Cornelis, H. Crews, O.F.X. Donard and P. Quevauviller, *Trace element speciation for environment, foods and health*, Royal Society of Chemistry, Cambridge, 2001.
45. R. Lobinski and F.C. Adams, *Spectrochim Acta, Part B.*, 1997, **52B**, 1865.

46. N.P. Vela and J.A. Caruso, *J. Biochem. Biophys. Methods*, 2000, **43**, 45.
47. C. Sarzanini and E. Mentasti, *J. Chromatogr, A.*, 1997, **789**, 301.
48. K.L. Sutton, R.M.C. Sutton and J.A. Caruso, *J. Chromatogr, A.*, 1997, **789**, 85.
49. G.K. Zoorob, J.W. McKiernan and J.A. Caruso, *Mikrochim. Acta*, 1998, **128**, 145.
50. N.H. Bings, J.M. Costa-Fernandez, J.P. Guzowski Jr., A.M. Leach and G.M. Hieftje, *Spectrochim. Acta B*, 2000, **55**, 767.
51. B. Bouyssiere, J. Szpunar and R. Lobinski, *Spectrochim. Acta*, 2002, **57B**, 805.
52. B. Michalke, *Trends Anal. Chem.*, 2002, **21**, 154.
53. B. Michalke, *Trends Anal. Chem.*, 2002, **21**, 142.
54. R.M. Barnes, *Fresenius' J. Anal. Chem.*, 1998, **361**, 246.
55. K. Takatera and T. Watanabe, *Anal. Chem.*, 1993, **65**, 3644.
56. J. Szpunar, *Trends Anal. Chem.*, 2000, **19**, 127.
57. J.L. Nielsen, A. Abildtrup, J. Christensen, P. Watson, A. Cox and C.W. McLeod, *Spectrochim. Acta, Part B.*, 1998, **53B**, 339.
58. S.F. Stone, R. Zeisler and G.E. Gordon, *Biol. Trace Element Res.*, **1990**, 85.
59. Z.B. Szokefalvi-Nagy, C. Bagyinka, I. Demeter, K.L. Kovacs, L.H. Quynh, *Biol. Trace Element. Res.*, **1990**, 93.
60. T.W.M. Fan, E. Pruszkowski and S. Shuttleworth, *J. Anal. At. Spectrom.*, 2002, **17**, 1621.
61. L. Dunemann and H. Reinecke, *Fresenius' J. Anal. Chem.*, 1989, **334**, 743.
62. B. Godlewska-Zylkiewicz, B. Lesniewska and A. Hulanicki, *Anal. Chim Acta.*, 1998, **358**, 185.
63. B. Michalke, *Fresenius' J. Anal. Chem.*, 1996, **354**, 557.
64. K.O. Olayinka, S.J. Haswell, and R. Grzeskowiak, *J. Anal. At. Spectrom.*, 1989, **4**, 171.
65. A.B. Soldado Cabezuelo, E. Blanco Gonzalez and A. Sanz Medel, *Analyst*, 1997, **122**, 573.
66. G.F. Van Landeghem, P.C. D'Haese, L.V. Lamberts and M.E. De Broe, *Anal. Chem.*, 1994, **66**, 216.
67. K. Wrobel, E. Blanco Gonzalez, K. Wrobel and A. Sanz Medel, *Analyst*, 1995, **120**, 809.
68. G. Alsing Pedersen and E.H. Larsen, *Fresenius' J. Anal.Chem.*, 1997, **358**, 591.
69. H. Emteborg, G. Bordin and A.R. Rodriguez, *Analyst*, 1998, **123**, 893.
70. P. Vinas, N. Campillo, I. Lopez Garcia and M. Hernandez Cordoba, *Chromatographia.*, 1996, **42**, 566.
71. P. Quevauviller, *Method Performance Studies for Speciation Analysis*, Royal Society of Chemistry, Cambridge, 1998.
72. S.A. Pergantis, K.A. Francesconi, W. Goessler and J.E. Thomas Oates, *Anal. Chem.*, 1997, **69**, 4931.
73. G. Zoorob, F. Byrdy Brown and J. Caruso, *J. Anal. At. Spectrom.*, 1997, **12**, 517.
74. H. Chassaigne and R. Lobinski, *Analysis*, 1997, **25**, M37.

75. G.R. Agnes and G. Horlick, *Appl. Spectrosc.*, 1994, **48**, 1347.
76. H. Chassaigne, V. Vacchina and R. Lobinski, *Trends Anal. Chem.*, 2000, **19**, 300.
77. S.A. Pergantis, W.R. Cullen and G.K. Eigendorf, *Biol. Mass Spectrom.*, 1994, **23**, 749.

CHAPTER 2

Element Specific Detection in Chromatography

1 Introduction

The beginnings of element specific detection in chromatography are associated with the use of flame AAS that was first employed as a GC detector for the speciation of alkyllead compounds.[1] Throughout the 70s and 80s improvements in GC-AAS interfaces consituted a major field of research in speciation analysis. In parallel, the use of flame photometric detection (FPD) with a bandpass filter selective to tin, electron capture detection (ECD), and especially electron impact MS were proposed by organic analytical chemists as alternative techniques for trace element speciation analyses. A considerable momentum to speciation research was given by the commercialisation in 1989 of an atmospheric pressure microwave induced plasma atomic emission spectrometer (MIP AES) dedicated entirely to element specific detection in GC.

For on-line HPLC detection, flame AAS was first applied in the end of the 70s despite the marked lack of sensitivity. It was not, however, until the mid 90s when wider availability of ICP MS made HPLC-ICP MS a common technique for speciation studies opening up the field of naturally existing metallobiomolecules (bioinorganic analytical chemistry). ICP MS has become a universal detector and allowed the employment of high resolution microsample separation techniques, such as capillary electrophoresis, for metal speciation analysis. Laser ablation ICP MS is becoming increasingly popular as a detector for 2D electrophoresis in selenoproteomics, to date a field reserved for detection by autoradiography or instrumental neutron activation spectrometry.

The wider use of ICP MS and the recent availability of a commercial interface have also contributed to the expansion of this technique for element specific detection in GC. This is happening despite the fact that the advantages of ICP MS over other GC detectors are not as well pronounced as for HPLC.

Because of the universality of ICP MS detection a separate section in this chapter is devoted to this technique. Separate chapters discuss in further detail the coupling of ICP MS with GC, HPLC and capillary electrophoresis.

2 Element Selective Detection in Gas Chromatography

Figure 2.1 shows an overview of element-specific detection techniques in gas chromatography.[2] The practical applications to volatile organometallic species are dominated by three techniques: GC-MIP atomic emission detection (AED), GC-ICP MS and GC-electron impact (EI) MS. The only exception is the determination of methylmercury in the environment where the position of GC-AAS and GC-atomic fluorescence spectrometry (AFS) is still remarkably strong. A similar statement applies, however to a lesser degree, to organotin analysis which can still be successfully performed by GC-flame photometric detector (FPD), especially when the improved pulsed flame photometric detector (PFPD) is used. MIP AES is preferred for the analysis of S, F and Cl of which ionization in an ICP is extremely difficult. The improved ion optics, and hence sensitivity of ICP MS spectrometers compensates for this lack, especially when a collision cell is used to improve selectivity.

Atomic Absorption Spectrometry

Flame AAS although used initially, was quickly abandoned because of insufficient sensitivity preventing applications to real life samples. An improvement in sensitivity required the avoidance of direct contact of the effluent with the flame gases and an increase in the residence time of the atoms in the light beam. An

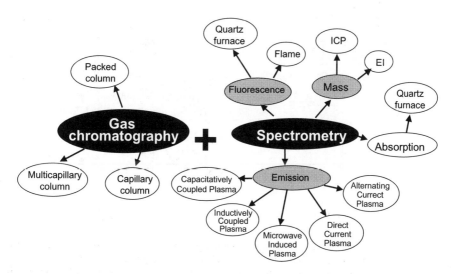

Figure 2.1 *Gas chromatography with element selective detection*

electrothermally heated silica tube was usually used as the atomisation cell. In the simplest arrangement (Figure 2.2a)[3] the chromatographic column was contained in the long arm of the T-furnace and the effluent then passes into the cross-piece atomiser purged with hydrogen and nitrogen (air) gases. The design was very simple and inexpensive but successful only for the determination of very volatile species, such as hydrides.

The interfacing of a quartz furnace to a capillary column for the analysis of less volatile species requires a more sophisticated design (Figure 2.2b).[4] Problems arise due to the condensation of higher boiling species on the cold surfaces induced by the addition of support gases or on cold spots created just in front of the quartz furnace. They can be avoided by heating the hydrogen and air prior to contact with analytes and the extension of the heating elements virtually into the furnace. Some designs allow the determination of very involatile species such as pentylated organotin compounds.[5]

Flame Photometric Detection

The GC effluent containing heteroatoms of interest is burned in a hydrogen-rich flame to produce molecular products that emit light. The emitted light is isolated from background emissions by a bandpass filter, detected by a photomultiplier and amplified. The selectivity of the FPD is limited by the light emissions of the continuous flame combustion products. Narrow bandpass filters limit the fraction of the element-specific light which reaches the detector but are not completely effective in eliminating flame background and hydrocarbon interferences.

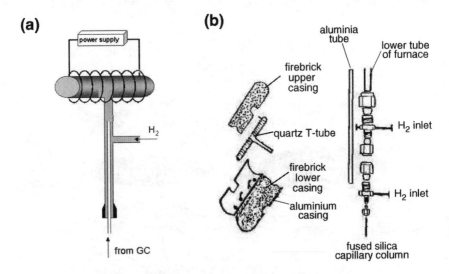

Figure 2.2 *Schemes of GC-QF AAS coupling designs: (a) packed column to quartz furnace, redrawn and adapted from ref. 3; (b) capillary column to quartz furnace* (b reprinted with permission from *Anal. Chem.*, 1985, **57**, 1299, copyright 1985, American Chemical Society)

Despite these problems FPD has enjoyed a strong position as a tin selective detector. Tin gives strong emission in a H_2-rich flame in the 360–490 nm (blue) and in the 600–640 nm (red) regions. A 600–610 nm bandpass filter is usually applied to avoid sulfur interference. The need for the chemical elimination of sulfur is frequently raised during the analysis of sediments in order to avoid the formation of artefacts.[6]

An advance in flame photometric detection is the pulsed flame photometric detector (PFPD) (Figure 2.3).[7] Unlike the traditional FPD which has a continuous flame, the PFPD is based on a pulsed flame for the generation of flame chemiluminescence. The detector operates with a fuel-rich mixture of hydrogen and air. This mixture is ignited and then propagates into a combustion chamber three to four times per second where the flame front extinguishes. Carbon light emissions and the emissions from the hydrogen/oxygen combustion flame are complete in two to three milliseconds, after which a number of heteroatomic species give delayed emissions which can last 4–20 ms. These delayed emissions are filtered with a wide bandpass filter, detected by an appropriate photomultiplier tube, and electronically gated to eliminate background carbon emission. The PFPD can detect at least 28 elements, 13 of them (S, P, N, As, Se, Sn, Ge, Ga, Sb, Te, Br, Cu and In) with high selectivity. The practical demonstration was limited to organotin; a 10-fold gain in sensitivity was claimed.[7]

(a)   **(b)**

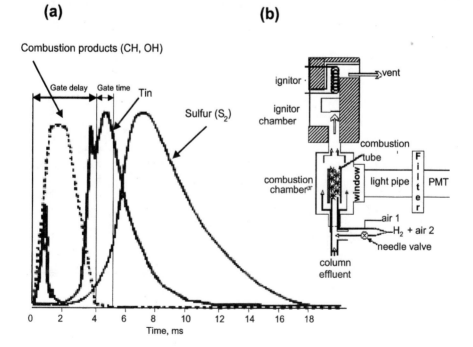

Figure 2.3 *Pulsed flame photometric detection in gas chromatography: (a) principle of operation, (b) scheme of the instrument*
(reprinted with permission from *Anal. Chem.*, 1995, **67**, 3305, copyright 1995, American Chemical Society)

Plasma Source Emission Spectrometric Detection

Plasmas compare favourably with both a chemical combustion flame and an electrothermal atomiser with respect to the efficiency of excitation of elements. The higher temperatures obtained in the plasma result in increased sensitivity and a larger number of elements can be efficiently determined. The use of different plasmas for element-specific detection in GC effluents has been critically reviewed.[8] The practical significance of these studies in terms of applications is practically non-existent with the exception of microwave induced plasma.

The coupling of GC–MIP AED has been extremely popular in speciation analysis of anthropogenic environmental contaminants and products of their degradation. This is due to its versatility and detection limits in the sub-picogram range which can be matched only by ICP MS.[8] Another factor contributing considerably to the popularity of GC–MIP AED has been the commercial availability of a purpose built integrated instrument since 1989. The system is based on the Beenaker TM_{010} cylindrical resonance cavity allowing the generation of helium plasma at atmospheric pressure, and a photodiode array detector enabling the simultaneous measurement of four emission wavelengths with a precision of 0.004 nm provided that they are not separated by more than 40 nm. The scheme of the instrument is shown in Figure 2.4.

GC–MIP AED offers sufficiently attractive analytical figures of merit to be applied on a routine basis to speciation of organotin and organolead compounds in the environment and methylmercury in biological tissues. It is being gradually

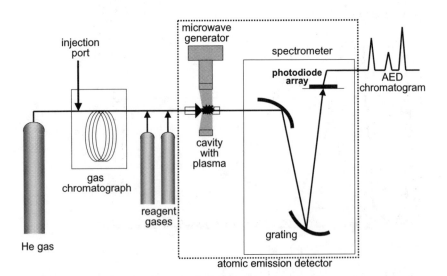

Figure 2.4 *Gas chromatography with MIP atomic emission detection: scheme of the instrumental setup*
(Agilent).

replaced by GC–ICP MS where lower detection limits allow a simpler sample preparation procedure, work with more dilute extracts, and especially a sensitive speciation analysis of complex organic matrices (*e.g.* gas condensates). The position of ICP MS has recently become stronger owing to the availability of a commercial interface.

Atomic Fluorescence Spectrometry

In atomic fluorescence spectrometry (AFS), ground state atoms created thermally, in a hydrogen flame or quartz furnace, are excited by a beam of light. However, instead of the absorbed light, as in AAS, the emission resulting from the decay of the excited states is measured (Figure 2.5a). AFS is a very specific element detection technique but, in order to be sensitive, a high output, high stability light source, *e.g.* a laser tunable to an element characteristic wavelength, is required. Line light sources, such as electrodeless discharge lamps are an alternative but their intensity, which determines the sensitivity, is usually too low to match the detection characteristics of AAS or ICP AES. The only exception is mercury for which a UV low-pressure mercury lamp has turned out to be an effective light source.[9]

An AFS detector coupled with a gas chromatograph is a commercially available hyphenated system allowing speciation of mercury (Figure 2.5b).[10] The GC effluent has to pass through a pyrolyser in order to form ground state atoms.[10] GC-AFS is a convenient method for mercury speciation in environmental matrices but the risk of artefacts due to the presence of hydrocarbons prevents this technique from successful application to more complex matrices.

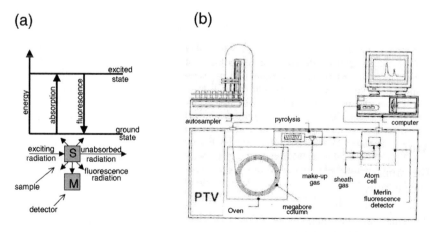

Figure 2.5 *Gas chromatography with atomic fluorescence spectrometric (AFS) detection: a) principle of the detection technique, b) scheme of the instrumental setup* (Reprinted from *Anal. Chim. Acta.*, 1999, **390**, 245, copyright 1999 with permission from Elsevier)

Electron Impact Mass Spectrometry

Mass spectrometry of molecular ions, which is a common detection technique in GC of organic compounds, is relatively seldom used in speciation analysis. Wide popularity is enjoyed by electron impact (GI) mass spectrometers owing to their availability as benchtop systems at relatively low cost. For quantitative analysis the systems are operated in the single ion monitoring mode which detection limits are two orders of magnitude lower than in the full scan mode used for structure elucidation. For most organometallic compounds detection limits at the low picogram level can be achieved in the single ion monitoring mode.

The unquestionable advantage of molecular ion MS is the confirmation of the identity of the detected compound. This is achieved by retention time matching. The drawback is the high background noise due to the ionisation of other (not necessarily metal-containing) species co-eluting with the analyte compound. Therefore when retention time standards are available, GC with MIP AES or ICP MS is preferred to EI MS. In the opposite case, *e.g.* the analysis of natural volatile metabolites of sulphur and selenium in garlic, molecular ion MS becomes indispensable for structure elucidation.[10a]

Because of its wide availability GC-MS is a recommended technique when standardisation of an analytical procedure is required. However, lower detection limits are usually obtained using element specific detectors such as MIP AES and ICP MS.

3   Element Selective Detection in HPLC

Several techniques have been used for the off-line determination of metals in fractions of the eluate collected from HPLC columns. They include instrumental neutron activation analysis (INAA),[11,12] total-reflection X-ray fluorescence (TXRF),[13,14] and especially graphite furnace (GF) AAS.[15] For on-line coupling ICP MS is the dominant detection technique. Some use of ICP AES is justified for the detection of sulfur and phosphorous. Atomic absorption and fluorescence spectrometries are used only in combination with post-column volatilisation but their practical significance for real life sample analyses is almost non-existent.

Atomic Absorption Spectrometry

The primary interest in HPLC-AAS lies in its simplicity, wide availability and compatibility with the mobile phases used in HPLC. AAS is not truly a multielement technique; nevertheless, with the latest instruments, up to four elements can be measured simultaneously, which is sufficient for practical applications. Flame AAS offers the possibility of an on-line approach. This technique is compatible both with the flow-rates and mobile phase composition (including organic solvents) commonly used in HPLC. Taking into account the widespread availability of this technique, it is no wonder that HPLC-AAS was the first hyphenated technique to be used for the determination of metal-protein complexes.[16]

Graphite furnace (GF) AAS analysis has been a sensitive technique for off-line analysis of metals in fractions collected after HPLC.[15,17] With an autosampler and flowthrough cells a high degree of automation can be achieved leading to a quasi on-line coupling.[18] The coupling of microbore AAS with GF AAS has been found to be promising.[19,20] The use of electrothermal (ET) AAS for metal speciation has been reviewed,[15] with particular attention to protein binding and trace element analysis in biological materials.[17]

The major fields of application include the detection of complexes with metals that give the most intense response in AAS (Cd, Zn, Cu) or of species that can be converted on-line into volatile hydrides (As, Se, Cd). Calibration graphs are reported to be rectilinear up to 20 μg mL^{-1} of metal.[21,22] The detection limits for As and Se after post-column hydride generation match those achieved by ICP MS with direct pneumatic nebulisation.

Inductively Coupled Plasma Atomic Emission Spectrometry

Some early work used a direct-currect plasma AES for the detection of serum proteins eluted from a size exclusion column[23] but the use of an ICP has quickly become well established. The HPLC-ICP AES coupling was the first hyphenated technique used for speciation of arsenic in biomaterials.[24] At present, the use of ICP AES is rapidly giving way to ICP MS.

ICP AES offers detection limits at the ng mL^{-1} level (in continuous infusion mode) which translates into 10–100 ng mL^{-1} for a transient signal of an analyte eluted from the column.[20,25–34] This can be lowered by the use of an ultrasonic nebuliser.[33] ICP AES offers, in comparison with ICP MS, a higher tolerance to matrix in terms of salt concentration (because of the absence of cones) and to organic solvents (because of higher power). Methanol or acetonitrile (>10%) can be readily tolerated. An advantage of ICP AES is the possibility of the simultaneous monitoring of sulfur or phosphorous with metals. A detection limit for S below 10 ng mL^{-1} is common. ICP AES instruments with axial plasmas seem to offer lower detection limits. Multielement capability is an advantage of instruments equipped with a polychromator. However, the sensitivity of ICP AES is generally inadequate to cope with the levels of most endogenous elemental species in real life samples.

Other Detectors

Some elements, *e.g.* Hg, As and Se can be successfully determined in HPLC effluents by quartz furnace (QF) AAS or AFS provided that they are converted post-column into volatile species, *e.g.* Hg° or hydrides that are swept into the atomiser. The post-column conversion into volatile species often involves UV photolysis or microwave-assisted digestion. A typical setup for HPLC with post-column volatilisation is shown in Figure 2.6.[35] In comparison with pneumatic

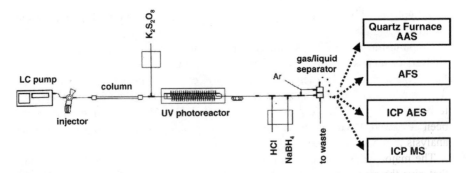

Figure 2.6 *Interface between HPLC and atomic spectrometry for post-column volatilisation,*
(adapted from *Trends Anal. Chem.*, 1995, **14**, 274, copyright 1995, with permission from Elsevier)

nebulisation a post-column hydride generation interface offers a 20–100-fold increase in sensitivity and elimination of interference from the sample matrix or mobile phase components. This is often off-set by background noise, especially in ICP MS, reducing the gain in detection limit to a factor of 2–10.[14]

4 ICP MS Detection in Chromatography and Electrophoresis

This section highlights the features of ICP MS that are common in its application as a detection technique. The interfaces between ICP MS and the particular separation techniques are discussed in dedicated chapters, 3, 4 and 5, for GC, HPLC and electrophoresis, respectively.

Principles of ICP MS

An ICP is usually an argon plasma formed in a quartz torch by coupling radio-frequency energy at 27.1 or 40 MHz through a load coil to form an oscillating magnetic field. The torch consists of a set of three concentric tubes. The nebuliser gas, which carries the analytes into the plasma, flows into the center tube, called the injector. The auxiliary gas flows around the injector tube and adjusts the horizontal position of the axial plasma relative to the torch. A third flow, coolant gas, flows tangentially through the outer tube, serving as the primary plasma gas, to cool the inside walls of the torch and to centre and stabilise the plasma. The temperature in the plasma can attain 8000 K and the energy is sufficient to ionise all but a few elements from the Periodic Table. The features of ICP MS have been exhaustively discussed.[36]

Liquid sample introduction is achieved *via* a nebuliser which creates an aerosol. A spray chamber is needed to separate any large droplets from the fine aerosol. An alternative is the use of a direct injection nebuliser DIN)[37] or DIHEN (direct injection high efficiency nebulizer)[36] The aerosol containing sample travels to the plasma where it is desolvated, vaporised, atomised, and ionised.[36] Dry aerosols from laser ablation or effluents from gas chromatography can be introduced directly through the injector tube but the risk of losses by condensation may be significant. In order to remedy this problem different interfaces have been developed (*cf.* Chapter 3).

Mass analysers employed for the separation of ions generated in an inductively coupled plasma typically include quadrupole, time-of-flight and sector-field instruments. The use of collision cells in front of the quadrupole instruments is becoming more and more widespread in order to eliminate polyatomic interference.[38] Multicollector instruments offer a higher precision than those equipped with a single collector, especially for the determination of isotope ratios in transient signals. A scheme of ICP MS using different mass analysers is shown in Figure 2.7.

ICP MS Using Quadrupole Analysers

A quadrupole mass filter consists of four parallel metal rods arranged in such a way that two opposite rods have an applied potential of $-(U + V\cos(\omega t))$ and the other two rods have potential of $-(U + V\cos(\omega t))$, where U is a dc voltage and $V\cos(\omega t)$ is an ac voltage. The applied voltages affect the trajectory of ions travelling down the flight path centered between the four rods. For given dc and ac voltages, only ions of a certain mass-to-charge ratio pass through the quadrupole filter and all other ions are thrown out of their original path.

The major advantages of a quadrupole analyser include its simplicity, relatively low cost and thus availability. The absolute detection limits are in the sub-picogram range although the latest generation instruments allow values down to 1 fg to be obtained for some elements. The unit resolution means that polyatomic interference can prevent the determination of some elements in some matrices, *e.g.* ^{75}As in sea water (interference from ^{40}Ar^{35}Cl). The overall performance is remarkably good, especially following chromatography where the interfering ions are often separated from the analyte species.

An emerging alternative to the analysis of elements plagued by polyatomic interferences (Fe, V, Cr, As, Se) is the use of hexapole collision or multipole dynamic reaction cells between the plasma and the quadrupole analyser.[38]

ICP Time-of-Flight Mass Spectrometry

Time-of-flight mass spectrometry uses the differences in transit time through a drift region to separate ions of different masses. It operates in a pulsed mode so ions must be produced or extracted in pulses. An electric field accelerates all ions into a field-free drift region wih a kinetic energy of qV, where q is the ion charge and V is the applied voltage. Since the kinetic energy of the ion is expressed as $0.5\ mv^2$, lighter ions have a higher velocity than heavier ones and reach the detector installed at the end of the drift region sooner.

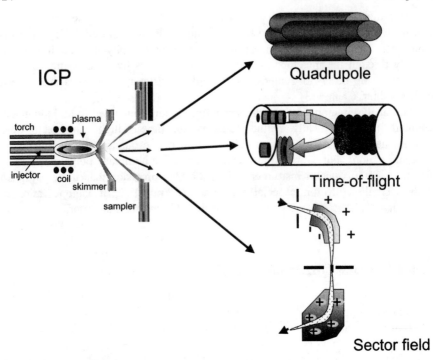

Figure 2.7 *A scheme of ICP with various mass analysers*

The ability to produce complete mass spectra at a high measurement frequency (typically $> 20\,000\ \text{s}^{-1}$) makes TOF MS nearly ideal for the detection of transient signals produced by high speed chromatographic techniques.[39] The measurement of a time-dependent, transient signal by sequential scanning using a quadrupole or a sector-field (single collector) mass spectrometer results in two major types of difficulties. The first is the limited number of isotope intensity measurements that can be carried out within the time-span of a chromatographic peak (especially from capillary or multicapillary GC). The other is the quantification error known as spectral skew, that arises during the measurement of adjacent mass-spectral peaks at different times along a transient signal.[40] Alleviating these difficulties requires an increase in the number of measurement points per time unit and simultaneous measurement of the isotopes the ratio of which is investigated. Figure 2.8 compares the peak definition in GC as a function of the speed of the mass detector used.[41]

The simultaneous extraction of all *m/z* ions for mass analysis in TOF MS eliminates the quantification errors of spectral skew, reduces multiplicative noise and makes TOF MS a valuable tool for determining multiple transient isotopic ratios.[40–43] ICP TOF MS suffers, however, from lower sensitivity in the monoelemental mode in comparison with the last generation of ICP quadrupole mass spectrometers. This loss of sensitivity is compensated by the fact that the number of isotopes determined during one chromatographic run is no longer limited by

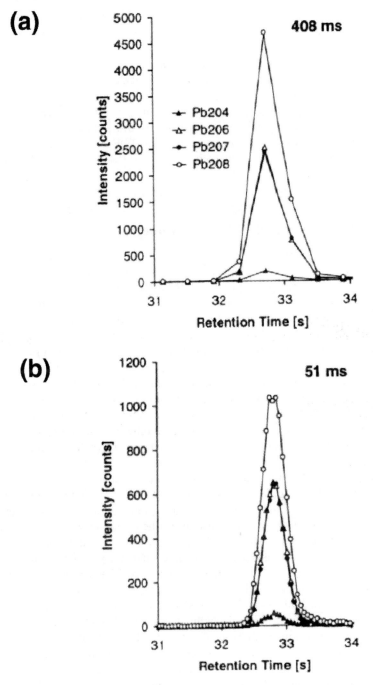

Figure 2.8 *Effect of the data acquisition speed on the peak definition in GC–ICP MS: a) with a quadrupole mass analyser, b) with a TOF mass analyser* (reprinted with permission from *Fresenius' J. Anal. Chem.*, 2001, **370**, 597, copyright 2001, Springer-Verlag)

peak definition (as in ICP MS employing a quadrupole analyser). The number of data points per chromatographic peak is independent of the number of measured isotopes.

Time-of-flight MS has been used in combination with GC, HPLC and capillary zone electrophoresis (CZE)[39,44] but its practical advantages over quadrupole MS for speciation analysis in real samples are far from being convincingly demonstrated. The need for the simultaneous detection of more than three isotopes in GC and of more than eight isotopes in HPLC is still scarce. The relatively low sensitivity prevents successful applications of ICP TOF MS in the analysis of real life samples by capillary electrophoresis.

ICP With Sector Field Mass Spectrometry

A high resolution instrument is based on the sequential focusing of an elemental ion using a magnetic sector and an electrostatic sector analyser. Consequently resolution can be increased to 10 000. Already a resolution of 3000 allows the elimination of many polyatomic interference. The increase in resolution brings a decrease in sensitivity but allows the easy determination of elements plagued by polyatomic interference. Another advantage is the reduction of the background noise because the photons formed in the plasma are prevented from reaching the detector, which results in a considerable decrease in the detection limits.

The detection limits can be further decreased by the use of sector field analysers. In the low resolution mode the sensitivity is about two orders of magnitude higher than that of a quadrupole spectrometer. Keeping blank levels low enough to exploit the extreme sensitivity becomes a major problem.[45] A wider expansion of high resolution ICP mass spectrometers (with potentially lower detection limits and larger freedom from interferences) is hampered by the prohibitive cost of instrumentation and the high maintenance costs.

Sector field MS has played an important role in the development of CZE-ICP MS coupling because of the low detection limits required.[46,47] It offers the possibility of the detection of sulfur useful for the determination of stoichiometry of metal-metallothionein complexes,[48] volatile sulfur compounds in bad breath[49] and phosphorus in DNA adducts.[50] A number of applications of size exclusion chromatography (SEC)-high resolution (HR) ICP MS have been shown but the advantages in comparison with the use of quadrupole analysers were unconvincing, the detection limits being controlled by contamination and the irreproducibility of chromatography of sub-picogram amounts of trace element complexes.

An interesting feature is the use of sector field instruments equipped with a multicollector detection facility for the accurate determination of element isotopic ratios in individual species.[51,52] A double-focusing instrument used for this purpose allowed a limit of 1 pg for ^{207}Pb (introduced as Et$_4$Pb).[51] Much lower detection limits were obtained by GC coupled to a single magnetic sector instrument equipped with a hexapole collision cell. A value of 2.9 fg was reported for the most abundant ^{208}Pb isotope.[52]

Laser Ablation ICP MS

The increased interest in the use of planar techniques for speciation studies, such as gel electrophoresis for the separation of seleno- and metalloproteins creates a demand for a suitable detector. In this context, laser ablation (LA) with its spatially resolved small spot sampling capacity is becoming an interesting sample introduction technique in combination with ICP MS.

A schematic diagram of a laser ablation system is shown in Figure 2.9. Usually a neodymium – yttrium aluminium garnet (Nd:YAG) laser operating in the UV ($\lambda = 266$ nm) or short UV ($\lambda = 213$ nm) range is used. The material vaporised by the laser pulse is swept into an ICP MS by a stream of argon. For gel electrophoresis detection a flat laser beam profile with an energy density of around 10 J cm^{-2} (at the sample surface) and diameter of 50 μm is used to ablate dry gel samples in a linear pattern through a determined (pre-stained) gel strip. The analysis is carried out with a speed of ca. 20 μm s^{-1} and a pulse frequency of 10 Hz.

To date, applications of LA ICP MS for monitoring trace elements separated in gels have been relatively scarce and include the detection of Co-binding serum proteins,[53] selenoproteins,[54] and Pb complexes with humic and fulvic acids.[55] The growth of this area is still hampered by the lack of sensitivity of LA ICP MS for the detection of very small metal amounts present in tiny spots. Recent advances in the simplification of the matrix by analysing electroblotted samples and custom-designing the geometry of laser ablation cells are likely to improve the analytical figures of merit of this technique.[56]

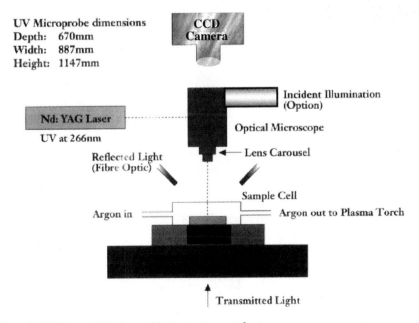

Figure 2.9 *Scheme of the laser ablation instrumental setup;* (adapted from a commercial leaflet Cetac)

References

1. B. Kölb, G. Kemmner, F.H. Schleser and E. Wiedeking, *Fresenius' Z. Anal. Chem.*, 1966, **221**, 166.
2. R. Lobinski, *Appl. Spectrosc.*, 1997, **51**, 260A.
3. Y.K. Chau, P.T.S. Wong and P.D. Goulden, *Anal. Chim. Acta*, 1976, **85**, 421.
4. D.S. Forsyth and W.D. Marshall, *Anal. Chem.*, 1985, **57**, 1299.
5. W. Dirkx, R. Lobinski, and F.C. Adams, *Anal. Sci.*, 1993, **9**, 273.
6. B. Lalere, J. Szpunar, H. Budzinski, P. Garrigues, and O.F.X. Donard, *Analyst*, 1995, **120**, 2665.
7. A. Amirav and H. Jing, *Anal. Chem.*, 1995, **67**, 3305.
8. R. Lobinski and F.C. Adams, *Spectrochim. Acta*, 1997, **52B**, 1865.
9. N. Bloom and W.F. Fitzgerald, *Anal. Chim. Acta.*, 1988, **208**, 151.
10. H.E.L. Armstrong, W.T. Corns, P.B. Stockwell, G. O'Connor, L. Ebdon and E. H. Evans, *Anal. Chim. Acta*, 1999, **390**, 245.
10a.E. Block, D. Putman and S.H. Zhao, *J. Agric. Food Chem.*, 1992, **40**, 2431.
11. C.K. Jayawickreme and A. Chatt, *J. Radioanal. Nucl. Chem.*, 1988, **124**, 257.
12. V.E. Negretti de Braetter, S. Recknagel and D. Gawlik, *Fresenius' J. Anal. Chem.*, 1995, **353**, 137.
13. K. Guenther and A. Von Bohlen, *Spectrochim. Acta*, 1991, **46B**, 1413.
14. K. Guenther, A. von Bohlen and C. Strompen, *Anal. Chim. Acta.* 1995, **309**, 327.
15. A.K. Das and R. Chakraborty, *Fresenius' J. Anal. Chem.*, 1997, **357**, 1.
16. K.T. Suzuki, *Anal. Biochem.*, 1980, **102**, 31.
17. P.C. d'Haese, G.F. Van Landeghem, L.V. Lamberts and M.E. De Broe, *Mikrochim. Acta.*, 1995, **120**, 83.
18. F. Laborda, M.V. Vicente, J.M. Mir and J.R. Castillo, *Fresenius' J. Anal. Chem.*, 1997, **357**, 837.
19. H. Emteborg, G. Bordin and A.R. Rodriguez, *Analyst*, 1998, **123**, 893.
20. H. Emteborg, G. Bordin and A.R. Rodriguez, *Analyst*, 1998, **123**, 245.
21. H. Van Beek and A.J. Baars, *J. Chromatogr.*, 1988, **442**, 345.
22. H. Van Beek and A.J. Baars, *At. Spectrosc.*, 1990, **11**, 70.
23. P.E. Gardiner, P. Braetter, V.E. Negretti and G. Schulze, *Spectrochim Acta*, 1983, **38B**, 427.
24. S. Kurosawa, K. Yasuda, M. Tagushi, S. Yamazaki, S. Toda, M. Morita, T. Uehiro and K. Fuwa, *Agric. Biol. Chem.*, 1980, **44**, 1993 .
25. W.A.J. De Waal, F.J.M.J. Maessen and J.C. Kraak, *J. Chromatogr.*, 1987, **407**, 253.
26. P.E. Gardiner, P. Braetter, B. Gercken and A. Tomiak, *J. Anal. At. Spectrom.*, 1987, **2**, 375.
27. K. Pomazal, C. Prohaska, I. Steffan, G. Reich and J.F.K. Huber, *Analyst*, 1999, **124**, 657.
28. H. Sunaga and K.T. Suzuki, *J. Liq. Chromatogr.*, 1988, **11**, 701.
29. K.T. Suzuki, *Analysis*, 1998, **26**, M57.
30. A. Mazzucotelli, A. Viarengo, L. Canesi, E. Ponzano and P. Rivaro, *Analyst*, 1991, **116**, 605.

31. A. Mazzucotelli, A. Viarengo, L. Canesi, F. De Paz, E. Ponzano and P. Rivaro, *Anal. Proc.*, 1991, **28**, 79.
32. A. Mazzucotelli and P. Rivaro, *Microchem. J.*, 1995, **51**, 231.
33. A. Mazzucotelli, V. Bavastello, E. Magi, P. Rivaro and C. Tomba, *Anal. Proc.*, 1995, **32**, 165.
34. P. Rivaro and R. Frache, *Analyst*, 1991, **122**, 1069.
35. R. Rubio, J. Alberti, A. Padro and G. Rauret, *Trends Anal. Chem.*, 1995, **14**, 274.
36. A. Montaser (Ed.), *Inductively coupled plasma mass spectrometry*, VCH, New York, 1998.
37. S.C.K. Shum and R.S. Houk, *Anal. Chem.*, 1993, **65**, 2972.
38. S.D. Tanner, V.I. Baranov and D.R. Bandura, *Spectrochim. Acta*, 2002, **B57**, 1361.
39. N.H. Bings, J.M. Costa-Fernandez, J.P. Guzowski Jr., A.M. Leach and G.M. Hieftje, *Spectrochim. Acta*, 2000, **55B**, 767.
40. A.M. Leach, M. Heisterkamp, F.C. Adams and G.M. Hieftje, *J. Anal. At. Spectrom.*, 2000, **15**, 151.
41. M. Heisterkamp and F.C. Adams, *Fresenius' J. Anal. Chem.*, 2001, **370**, 597.
42. K. Haas and J. Feldmann, *Anal. Chem.*, 2000, **72**, 4205.
43. J.R. Baena, M. Gallego, M. Valcarcel, J. Leenaers and F.C. Adams, *Anal. Chem.*, 2001, **73**, 3927.
44. M. Balcerzak, *Anal. Sci.*, 2003, **19**, 979.
45. J.M. Gonzalez LaFuente, J.M. Marchante-Gayon, M.L. Fernandez Sanchez, A. Sanz-Medel, *Talanta*, 1999, **50**, 207.
46. A. Prange and D. Schaumlöffel, *J. Anal. At. Spectrom.*, 1999, **14**, 1329.
47. A. Prange, D. Schaumlöffel, P. Brätter, A. Richarz and C. Wolf, *Fresenius J. Anal. Chem.*, 2001, **371**, 764.
48. D. Schaumloeffel, A. Prange, G. Marx, K.G. Heumann and P. Braetter, *Anal. Bioanal. Chem.*, 2002, **372**, 155.
49. J. Rodriguez-Fernandez, M. Montes-Bayon, R. Pereiro and A. Sanz-Medel, *J. Anal. At. Spectrom.*, 2001, **16**, 1051.
50. C. Siethoff, I. Feldmann, N. Jakubowski and M. Linscheid, *J. Mass Spectrom.*, 1999, **34**, 421.
51. E.M. Krupp, C. Pécheyran, H. Pinaly, M. Motelica-Heino, D. Koller, S.M.M. Young, I.B. Brenner and O.F.X. Donard, *Spectrochim. Acta*, 2001, **56B**, 1233.
52. E.M. Krupp, C. Pécheyran, S. Meffan-Main and O.F.X. Donard, *Fresenius J. Anal. Chem.*, 2001, **370**, 573.
53. J.L. Nielsen, A. Abildtrup, J. Christensen, P. Watson, A. Cox and C.W. McLeod, *Spectrochim. Acta*, 1998, **53B**, 339.
54. T.W.M. Fan, E. Pruszkowski and S. Shuttleworth, *J. Anal. At. Spectrom.*, 2002, **17**, 1621.
55. R.D. Evans and J.Y. Villeneuve, *J. Anal. At. Spectrom.*, 2000, **15**, 157.
56. M. Wind, I. Feldmann, N. Jakubowski, W.D. Lehmann, *Electrophoresis*, 2003, **24**, 1276.

CHAPTER 3

Gas Chromatography with ICP MS Detection

1 Introduction

Gas chromatography is an ideal sample introduction technique for atomic spectroscopy because of the absence of a condensed mobile phase and the very high separation efficiency (number of theoretical plates). However, it is limited solely to organometallic compounds which are volatile and thermally stable in the native form or which can be converted to a gas chromatographable form by means of derivatisation. Avoiding thermal degradation of the analyte species in the injector and condensation at the interface during transport from the end of the column to the atomisation/excitation/ionisation source are major challenges to efficient interfacing of GC with atomic spectroscopy. The elemental species determined preferably by GC have been summarised in Figure 3.1. These are mostly organometallic anthropogenic contaminants and products of their degradation or environmental transformation.

ICP MS is rapidly conquering the field of element-selective detection in GC despite the presence of a number of cheaper, sometimes even better performing but less versatile competitive techniques. Pioneering work on GC-ICP MS coupling goes back to the mid 80s with the landmark papers of Van Loon et al.[1] and Chong and Houk.[2] Since then ICP quadrupole MS has been undergoing constant improvement leading to the wider availability of more sensitive, less interference prone, smaller and cheaper instruments which favour their use as chromatographic detectors. The combination of capillary GC with ICP MS has become an ideal methodology for speciation analysis of organometallic compounds in complex environmental and industrial samples because of the high resolving power of GC and the sensitivity and specificity of ICP MS.[3-7] Indeed, features such as low detection limits reaching the one femtogram (1 fg) level, high matrix tolerance allowing the direct analysis of complex samples, and isotope ratio measurement capacity enabling accurate quantification by isotope dilution position ICP MS at the head of all GC element specific detectors.

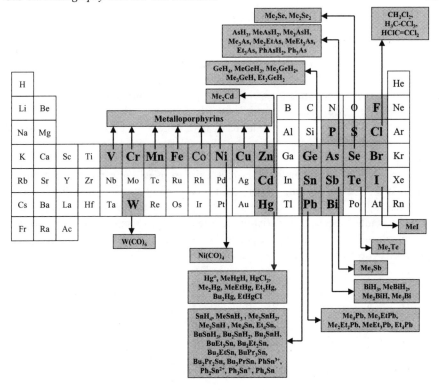

Figure 3.1 *The elemental species determined by GC with element selective detection* (from ref. 7)

The introduction of ICP time-of-flight MS increased the speed of data acquisition allowing multiisotope measurement of millisecond-wide chromatographic peaks and improving the precision of isotope ratio determination.[8–12] Even better precision was announced in preliminary reports on the use of sector field multicollector instruments.[13,14] These instrumental developments have gone in parallel with the miniaturisation of GC hardware allowing the time-resolved introduction of gaseous analytes into an ICP, *e.g.* based on microcolumn multicapillary GC, and sample preparation methods including microwave-assisted, solid-phase microextraction or purge-and-capillary-trap automated sample introduction systems.[15] To date, GC-ICP MS is virtually the only technique capable of direct speciation analysis of As and Hg in natural gases and gas condensates at the $\mu g\,L^{-1}$ level (*cf.* Chapter 12).[16] A GC-ICP MS interface is commercially available.

2 Derivatisation Techniques for Gas Chromatography of Organometallic Species

A number of native organometallic compounds are volatile enough to be separated by GC. They include tetraalkyllead species ($Me_nEt_{4-n}Pb$) (n = 1 ÷ 4), methylselenium compounds (*e.g.* Me_2Se, Me_2Se_2), some organomercury compounds ($MeHg^+$, Me_2Hg) as well as naturally occurring metalloporphyrins. They can either be readily purged with an inert gas or extracted into a non-polar solvent and subsequently chromatographed by thermal desorption, packed column or capillary GC.

The majority of organometallic species exist in quasi-ionic polar forms which have relatively high boiling points and often poor thermal stability. To be amenable to GC separation they must be converted to non-polar, volatile and thermally stable species. The derivative chosen needs to retain the structure of the element-carbon bonds to ensure that the identity of the original moiety is conserved. The most common derivatisation methods include:

(1) conversion of inorganic and small organometallic ions into volatile covalent compounds (hydrides, fully ethylated species) in aqueous media;

(2) conversion of larger alkylmetal cations *e.g.* $R_nPb^{(4-n)+}$ to saturated non-polar species using Grignard reagents; and

(3) conversion of ionic species to fairly volatile chelates (*e.g.* dithiocarbamate, trifluoroacetone) or other compounds.

The three methods are fairly versatile in terms of the organometallic species to be derivatised and the choice depends on the concentration, the matrix and the sample throughput required. Frequently, the derivatives are concentrated by cryotrapping or extraction into an organic solvent prior to injection onto a GC column.

Derivatisation (chemical modification) techniques for GC in speciation analysis have been reviewed.[17]

Derivatisation by Hydride Generation

Several elements (Hg, Ge, Sn, Pb, Se, Te, Sb, As, Bi and Cd) can be transformed into volatile hydrides, forming the basis of their determination.[18,19] The usefulness of this procedure for speciation analysis, however, is severely restricted by either the thermodynamic inability of some species to form hydrides, or by considerable kinetic limitations to hydride formation. Nevertheless, the technique is still essential for some classes of compounds.

Selenium, As and Sb readily form hydrides only when present in their lower oxidation states; the higher states need to be reduced beforehand. Thus, all inorganic species of these elements eventually form the same hydride (SeH_2, AsH_3 and SbH_3, respectively) precluding simultaneous chromatographic speciation analysis. Methyl- and dimethylarsonic (or stibonic) acid can be discriminated in one GC run on hydride generation by the production of volatile $MeAsH_2$ and Me_2AsH (or $MeSbH_2$ and Me_2SbH) respectively. Trialkyllead species form stable hydrides whereas dialkylleads are non-reactive. Mercury(II) and methylmercury, as well as germanium and methylgermanium species[18] can be converted to gas chromatographable hydrides. Hydride generation has attracted the most interest for organotin speciation analysis because of its capability for the simultaneous determination of ionic methyl and butyl species in one chromatographic run.[20,21] Hydride generation with $NaBH_4$ is prone to interference with transition metals which affect the reaction rate and analytical precision.[19] Sources of error in derivatisation of organometal(loid) species by $NaBH_4$[22] and artefact formation during hydride generation from antimony species have been discussed.[23]

Derivatisation with Tetraalkyl(aryl)borates

The vulnerability of hydride generation to interference in real samples and the restricted versatility can, to a certain degree, be overcome by replacing $NaBH_4$ by alkylborates. The most common derivatisation procedures rely on ethylation with sodium tetraethylborate ($NaBEt_4$) which is water soluble and fairly stable in aqueous media.[24]

Attempts at multielement and multispecies derivatisation using $NaBEt_4$ have been discussed.[25] Methyl-, butyl- and phenyltin compounds react readily with $NaBEt_4$ to form thermally stable gas chromatographable species. Whereas methyltins can be purged upon derivatisation, other species need to be extracted because of their poorer volatility. All alkyllead species also react readily but only methyllead species can be unambiguously discriminated, as the derivatisation of ethyl- and inorganic lead will lead to the formation of the same product – $PbEt_4$. Methylmercury and inorganic mercury can be speciated in one run upon purge-and-trap preconcentration. Selenite can be determined selectively and free of interference after derivatisation with $NaBEt_4$.[26]

Sodium tetrapropylborate has been introduced to allow discrimination between ionic ethyllead and inorganic lead species after derivatisation.[11,27] Tetramethylammonium tetrabutylborate has been proposed for the same purpose.[28] The possibility of the simultaneous determination of Sn, Hg and Pb following propylation with $NaBPr_4$ has been demonstrated.[27] *In situ* phenylation using sodium tetraphenyl borate has been studied with some success.[29]

Derivatisation with Grignard Reagents

An alternative to hydride generation or ethylation with alkylborates is derivatisation with Grignard reagents. This method, recommended especially in the case of complex matrices, is fairly versatile but requires an aqueous-free medium for the reaction to be carried out. In practice, it is applicable to extracts containing complexes of an organometallic compound with dithizone, dithiocarbamates or tropolone. Whereas Grignard derivatisation still remains the primary method for lead speciation analysis,[30–33] its position has been eroded for organotin speciation in favour of the less cumbersome and time-consuming derivatisation with NaBEt$_4$.[34] Other applications of Grignard reagents *e.g.* for the derivatisation of organomercury, have been limited to one research group.[35,36] An interesting curiosity is direct Grignard pentylation, demonstrated for an organotin-contaminated lard sample.[37]

Grignard reagents proposed for derivatisation in speciation analysis by GC with plasma source spectrometric detection have included: methyl-, ethyl-, propyl-, butyl- and pentylmagnesium chlorides or bromides. Lower-alkyl magnesium salts are generally preferred due to the smaller molecular mass and, hence, the higher volatility of the resulting species which makes the GC separation faster with less of the column carryover problems associated with derivatised inorganic forms (which are often present in large excess). In addition, the baseline is more stable and fewer Grignard reagent-related artefacts occur. Conversely, the low volatility of species derivatised by pentylmagnesium chloride may facilitate concentration by evaporation and, hence, allow a more efficient enrichment.

The unreacted Grignard reagent needs to be destroyed prior to injection of the derivatised extract onto a column, which is achieved by shaking the organic phase with dilute H$_2$SO$_4$. As a final step of the procedure, the organic phase is dried, for example, over anhydrous Na$_2$SO$_4$ and injected onto the GC column.

Other Derivatisation Techniques

The formation of volatile acetonates, trifluoroacetonates and dithiocarbamates is a popular derivatisation technique for inorganic trace element analysis by GC.[38] Kinetic restrictions or the thermodynamic inability of many species to react, and small differences in retention times for derivatised species of the same element make chelating agents of limited importance as derivatisation reagents for speciation analysis.

Selenoamino acids have been derivatised using isopropylchloroformate and bis(*p*-methoxyphenyl) selenoxide,[39] pyridine and ethyl chloroformate[40] or silylated with bis(trimethylsilyl)acetamide.[41] Selenomethionine forms volatile methylselenocyanide with CNBr.[42,43] Selective determination of Se(IV) and Se(VI) using GC-FPD and GC-MS after derivatisation of Se(IV) with 4,5-dichloro -1,2-phenylenediamine has been reported.[44]

Methods for the conversion of arsenic compounds to volatile and stable derivatives are based on the reaction of monomethylarsinic acid and dimethylarsonic acid with thioglycolic acid methyl ester (TGM).[45,46]

3 Separation of Organometallic Species by GC

The mobile phase in GC is usually helium which not only enables the quantitative transport of the sample to the detector without nebulisation, but also accounts for a low background contribution in the detector itself. GC techniques can be divided into different categories according to:

(1) the choice of the column used – packed, capillary, high temperature capillary, multicapillary;
(2) the sample introduction principle – direct injection (split, splitless, on-column) of an organic solution, solid-phase microextraction (SPME), headspace-injection or purge-and-trap using cryofocussing.

Choice of the Column

Packed column GC was used in early elemental speciation studies. Packed columns can, by design, handle high flow rates and large sample sizes, but the efficiency and resolution are compromised because of the high dispersion of analytes on the column. Packed column GC-ICP MS is a favorable technique to follow hydride generation purge-and-trap because of the easier handling of highly volatile species at temperatures below $-100\,^{\circ}\text{C}$.[47,48]

Capillary GC offers improved resolving power over packed column GC which is of importance for separation of the complex mixtures of organometallic compounds found in many environmental samples. The reduced sample size and high dilution factor caused by the make-up gas necessary to match the spectrometer's optimum flow rate, result in a loss of sensitivity. The gain in terms of resolution among packed, wide-bore and capillary GC columns for the same type of application (each time with optimised separation conditions) is shown in Figure 3.2.[49]

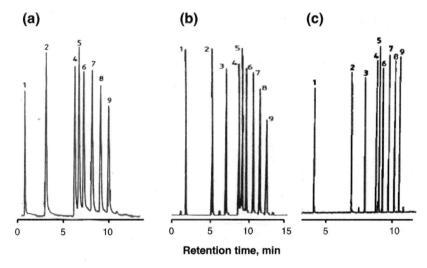

Figure 3.2 *Comparison of separation capabilities of different GC chromatographic columns; chromatograms obtained for a mixture of organotin species: 1–Me$_3$SnPe, 2–Me$_2$SnPe$_2$, 3–Pr$_3$SnPe, 4–Bu$_4$Sn, 5–MeSnPe$_3$, 6–Bu$_3$SnPe, 7–Bu$_2$SnPe$_2$, 8–BuSnPe$_3$ (internal standard, not added for packed column GC-AAS), 9–Pe$_4$Sn; (a) packed column GC-AAS, (b) megabore column GC-AAS, (c) capillary column GC-AAS*
(Reprinted with permission from *Anal. Sci.*, 1993, **9**, 273)

Recently, a number of papers have appeared on rapid (flash) GC employing columns that consist of a bundle of 900–2000 capillaries of a small (20–40 μm) internal diameter, referred to as multicapillary columns[50] (Figure 3.3a). Such a bundle allows the elimination of the deficiencies associated with the use of capillary and packed columns while preserving the advantages of both. Separations by multicapillary GC are carried out at high flow rates which minimises the

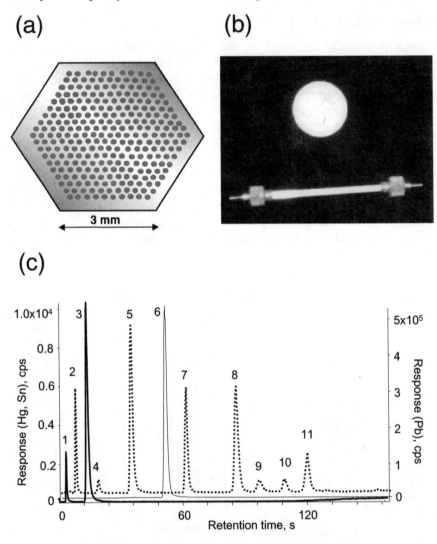

Figure 3.3 *The potential of multicapillary microcolumns for rapid multielement speciation analysis: (a) cross-section of a column; b) photograph of a multicapillary microcolumn in comparison with a US 10 cents coin; c) a chromatogram: 1– MeEtHg, 3–Et₂Hg, 5–Et₄Sn, 6–Et₄Pb, 7–BuSnEt₃, 8–Bu₂SnEt₂, 11–Bu₃SnEt, 2,4,9,10–unidentified (thick line – Hg, dotted line – Sn, regular line – Pb)* (reprinted with permission from *Anal. Chem.*, 1999, **71**, 4534, copyright 1999, American Chemical Society)

dilution factor and facilitates the transport of the analytes to the plasma. The coupling between multicapillary (MC) GC and ICP MS offers 0.08 pg detection limits for Hg speciation.[51]

An interesting feature is the use of multicapillary microcolumns for sample introduction into an ICP. Rodriguez et al.[52] have shown isothermal separations of organometallic species using a 50 mm column (Figure 3.3b) which opens the way to the miniaturisation of GC sample introduction units, possibly making the classical GC oven redundant. A chromatogram (Figure 3.3c) demonstrates the potential of this technique for rapid multielement speciation analysis.[53]

Gas chromatography using chiral stationary phases, such as L-valine-tert-butylamide[54] has been shown to resolve N-trifluoroacetate-O-isopropyl derivatives of D- and L-selenomethionine enantiomers in selenium nutritional supplements.[54]

Purge-and-Trap Using Cryofocussing

A typical instrumental setup consists of a reaction vessel, where the sample is brought in contact with a reagent (usually $NaBH_4$ or $NaBEt_4$) and a U-tube filled with a GC stationary phase maintained in liquid nitrogen, where the hydrides or ethyl derivatives are trapped (Figure 3.4a).[55] A water trap (another U-tube maintained at a subfreezing temperature, ca. $-15\,°C$) is often installed upstream from the sorbent tube to avoid its blocking with ice crystals. Purge-and-trap setups enjoyed a lot of interest in early speciation research because of the large preconcentration factors achieved and hence the feasibility of speciation analysis with relatively insensitive detectors, such as AAS. A surge of interest was brought about by the availability of commercial purge-and-trap injectors mountable on capillary GC (Figure 3.4b).[25]

A semi-automated compact interface for the time-resolved introduction of gaseous analytes from aqueous solutions into an ICP MS without the need for a full size GC oven has been described.[56] The working principle (Figure 3.4c) was based on purging the gaseous analytes with an inert gas, drying the gas stream using a 30 cm tubular Nafion membrane and trapping the compounds in a thick film-coated capillary tube followed by their isothermal separation on a multi-capillary column.[57] Recoveries were reported to be quantitative up to a sample volume of 50 mL.[56–58]

Cryotrapping methods are widely used in speciation analysis of air samples (cf. Chapter 8) where they are beyond any competition. The increasing availability of more sensitive ICP MS results in the choice of solvent extraction followed by injection on a GC column for other samples. Purge-and-cryotrap-QF AAS and -atomic fluorescence spectrometry (AFS) methods still remain a valid alternative for speciation analysis of mercury.

Solid-Phase Micro-Extraction

Solid-phase micro-extraction (SPME) is a solvent-free technique which offers numerous benefits such as simplicity, reduced sample amount used for analysis,

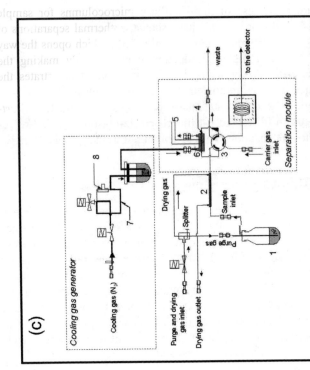

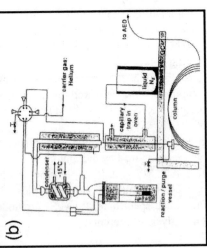

Figure 3.4 *A typical purge-and-trap setup: a) early purge-and-trap using a filled U-tube; b) purge-and-and-capillary-trap injector mounted on a GC (from Ref. 25); c) purge-and-capillary-crytrap using a Nafion drier (1 – purge vessel, 2 – Nafion dryer, 3 – 6-port valve, 4 – capilliary cryotrap, 5 – trap heater (SS tube), 6 – cooling chamber, 7 – copper tube (¼"), 8 – adjustable nitrogen restrictor), redrawn and adapted from ref. 57*

*(a reprinted with permission from Anal. Chem., 1986, **58**, 35, copyright 1986, American Chemical Society)*

low cost and the compatibility with on-line analytical procedures. SPME is based on an equilibrium between the analyte concentrations in the headspace and in the solid-phase fibre coating. Low extraction efficiencies are sufficient for quantification but the amount of the analyte retained on the fibre may be very small, hence the interest in the combination of the high sensitivity of ICP MS with SPME-GC. The combination SPME-GC-ICP MS is an emerging analytical tool for elemental speciation in environmental and biological samples.[59]

SPME-GC-ICP MS was first used by Moens *et al.*[60] for the simultaneous speciation of organomercury, -lead and -tin compounds ethylated *in situ* with NaBEt₄. Headspace SPME at non-equilibrium conditions was optimised as an extraction/preconcentration method for triphenyltin, derivatised using NaBEt₄ in tetramethylammonium hydroxide (TMAH) and KOH-EtOH extracts of potato and mussel samples.[61] Direct SPME (from the aqueous phase) was studied but the sensitivity was an order of magnitude lower. Detection limits are at the pg L^{-1} level.[61,62] The commbination SPME-GC-FPD is schematically shown in Figure 3.5 together with a typical chromatogram.[63]

Similar in principle to SPME is an extraction technique using stir bars coated with a relatively thick (0.3–1 mm) layer of poly(dimethylsiloxane): stir bar sorptive extraction (SBSE).[64] The stir bar (1–4 cm in length) is added to an aqueous sample and stirred. After a certain time, the bar is removed and thermally desorbed into a GC. Owing to the much larger volume of the stationary phase the extraction efficiency in SBSE is by far superior to that of SPME. The instrumental

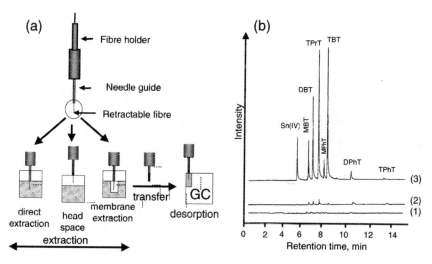

Figure 3.5 *(a) Principle of SPME; and (b) a typical chromatogram obtained by SPME-GC-FPD illustrating the preconcentration factor (1 – extraction into isooctane, 2 – SPME-magnetic stirring, 3 – SPME-mechanical stirring); MBT – monobutyltin, DBT – dibutyltin, TPrT – tripropyltin (internal standard), MPhT – monophenyltin, TBT – tributyltin, DPhT – diphenyltin, TPhT – triphenyltin (100 pg ml⁻¹ each)*
(from Ref. 63)

detection limits reported for organotin compounds were fabulous ($10 \, \mathrm{fg \, L^{-1}}$), in practice, however, values of $0.1 \, \mathrm{pg \, L^{-1}}$ could be achieved.[65]

Large Volume Injection in GC

The major disadvantage of capillary columns is their restricted capacity. Typically only 1–2 μL of the analyte solution can be injected without column overloading. This is often only a fraction of the available extract volume. Therefore, the performance of capillary GC is degraded in comparison with purge-and-trap systems which detect practically all the analyte available in a relatively large (10–50 mL) sample aliquot. Advanced systems equipped with electronic pressure control allow injection of up to several hundred microlitres without noticable peak shape deterioration. However, in many cases the large solvent load fouls the detector, resulting in the eventual degradation of the analytical performance of the system.

In an alternative approach the solvent is removed on-line prior to transfer of the analytes onto the column. The principle of preconcentration is based on the differences in volatility of the solvent and is shown in Figure 3.6.[66] A solution of

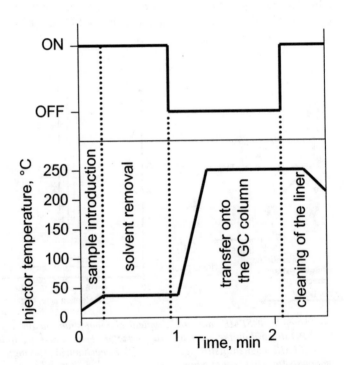

Figure 3.6 *Principle of in-liner preconcentration prior to GC with element selective detection. Injection system: KAS 503 PTV (Gerstel, Germany) (from Ref. 66)*

fairly involatile analytes in a volatile solvent is injected onto a cooled (0–10 °C) Tenax packed liner through which a stream of He gas is passing, sweeping the solvent vapour off the column. Once the bulk of the solvent has been removed the He carrier gas is switched to flow through the column while the temperature of the liner is raised to the effective injection temperature, prompting the release of the analytes.

4 Interfacing GC to ICP MS

The basic requirement for an interface is that the analytes should be maintained in the gaseous form during transport from the GC column to the ICP, in such a way that any condensation is prevented. This can be achieved either by an efficient heating of the transfer line avoiding cold spots, or by using an aerosol carrier. This results in two basic types of design for the GC-ICP MS interface:

(1) based on direct connection of the transfer line to the torch (Figure 3.7).[52] In this type of interface the spray chamber is removed and the transfer line inserted part of the way up the central channel of the torch. The required efficiency of heating depends on the species to be analysed.
(2) based on mixing the GC effluent with an aqueous aerosol in the spray chamber prior to introduction into the plasma (Figure 3.8).[14]

Regardless of the type of interface the addition of oxygen to the plasma gas is essential to prevent carbon deposition (and sometimes metal entrapment) and to reduce the solvent peak. Interface designs have been reviewed in detail.[7]

Interface Designs Based on Direct Connection of Transfer line to Torch

A GC effluent at a flow rate of a few mL min^{-1} requires a make up carrier gas to attain a sufficient flow to get the analytes into the central channel of the plasma. Plasma operating conditions are optimised by Xe gas added *via* a T-piece at a constant mass flow rate to the Ar nebuliser gas using a mass flow controller. The Xe the intensity of which is measured during the chromatography also serves as an internal standard. The benefit is the absence of an aerosol in the plasma which limits energy losses for desolvation and vaporisation, results in high sensitivity and reduces polyatomic interference. The basic features differentiating the interface designs within this type are whether and how the heating of the transfer line is accomplished and how easily the coupling and decoupling can be realised.

In the simplest case, volatile species cryotrapped and released by thermal desorption, a simple glass piece, connected to the end of the U-shaped glass chromatographic column, assures the transport of methyl and ethyl species of Hg, Pb and Sn, As and Ge.[48,67] A similar design has been used for the transfer of butyltin hydrides thermally desorbed from a packed column.[68] A high velocity Ar carrier make up gas flow was added post-column to prevent condensation of the

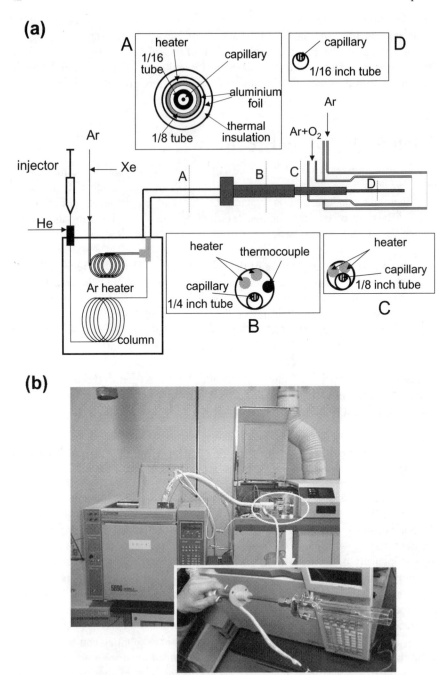

Figure 3.7 *(a) Design; and (b) a photograph of a heated GC-ICP MS interface; A, B, C, D insets present the respective cross-sections*
(Reproduced with permission from *Anal. Chem.*, 1999, **71**, 4534, copyright 1999, American Chemical Society)

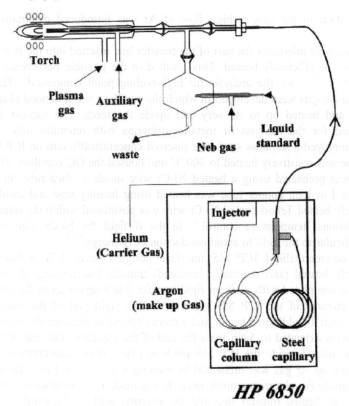

Figure 3.8 *Design of a GC-ICP MS interface via a spray chamber*
(reprinted with permission from *Spectrochim. Acta*, 2001, **56B**, 1233, copyright
2001, Elsevier)

tin species in the transfer line. Peak profiles for mono- and dibutyl hydrides were
satisfactory but pronounced tailing was observed for tributyltin hydride.[68] A
heating block at the exit from the oven (at the mixing point with the make up
argon) was necessary. In another design a multicapillary column has been
interfaced to ICP MS for the analysis of ethylmercury derivatives, upon
ethylation, liquid-gas extraction and capillary cryotrapping.[51,56] This design
produced negligible peak broadening for the ethylated mercury species; less
volatile compounds than Et_2Hg produced broad peaks inappropriate for quantification.

In order to improve the transport of less volatile analytes to the plasma, heated
interfaces have been designed. The use of a 60 cm long Al rod, wound in a
heating tape and housing the capillary in a longitudinal slot has been
proposed.[69–72] Smaele *et al.* have used a resistively heated stainless tubing.[73] Ar
gas was made to flow around the transfer fused silica capillary in the heated
transfer line. In another design a heated block was placed at the exit of the gas
chromatograph.[74,75] The transport of the analytes to the plasma was achieved with
a flexible 1.5 mm i.d. PTFE 80 cm tube at room temperature. In order to prevent

condensation in the tube a high flow of Ar was introduced externally to the column.

In the above interfaces the part of the transfer line inserted into the ICP injector could not be efficiently heated. This resulted in the possible occurrence of cold spots, especially for the analysis for high-boiling point compounds. Therefore, interface designs were developed in which the capillary was extended close to the plasma and heated up to its very end inside the torch. The earliest design,[76] developed for the analysis of metalloporphyrins with retention indices above 6000, employed a stainless steel tube inserted concentrically into an ICP injector. The tube was resistively heated to 400 °C and housed the GC capillary. The make up Ar was preheated using a heated Ni-Cr wire inside a silica tube. In another design a 1 m long copper tube was heated using heating tape and insulated. A resistively heated Teflon-coated Ni-Cr wire was positioned within the transfer line for additional temperature control.[10] In the rf field the Ni-Cr wire was kept perpendicular to the field to avoid conduction of rf energy.[77,78]

The commercialised ICP MS interface design (Figure 3.7) is based on a resistively heated (and thermally insulated) transfer line housing the capillary. This was completed with a 10 cm rigid transfer line inserted up to the end of the central channel of the ICP MS torch.[52,79–82] The rigid part of the transfer line encapsulated an additional heater and a thermocouple to measure the temperature. The heaters extended to 5 cm from the end of the capillary. The end of the rigid part was placed inside the ICP MS torch in place of the conventional injector. The make up Ar gas was preheated by passing it through a 1 m 1/16 inch coil placed inside the chromatographic oven. It was made to flow between the internal wall of the heated transfer line and the external wall of the capillary tube to merge with the GC carrier gas just before the plasma.[52]

Interface Designs via a Spray Chamber

The major inconvenience of the transfer line-to-torch interface is the need to dismantle and reassemble the system when the instrument is also used for routine elemental analyses of solutions. The conditions in the plasma are readily affected by changes in the composition of the matrix exiting the chromatographic column. An internal standard, *e.g.* Xe gas, added to the make up Ar is required to correct for the varying ionisation conditions.[52,80,81] However, correction in this way for mass fractionation in isotope dilution GC-ICP MS analysis is difficult. In order to alleviate the dis- and reassembling problems an interface was designed which allowed the selection of either the GC effluent or the sample aerosol using a 'zero dead volume valve' and required no system reconfiguration when switching between the two modes.[83,84] The make up gas was added using a sheathing device in such a way that it surrounded the sample (rather then being completely mixed with it), maintaining the analyte in the central channel of the plasma.[83,84]

Feldmann *et al.* proposed a design in which the GC effluent was mixed with an aqueous aerosol, ensuring stable plasma conditions.[85] The transfer capillary was inserted into the torch injector and an aqueous solution was continuously aspirated into the plasma *via* a T-piece connection using a conventional Meinhard nebuliser *via* a Scott double-pass spray chamber. In this way, the aerosol was transfered into the plasma together with the analyte gas. In a modified design a low-volume water-cooled cyclonic spray chamber was used.[14] An internal standard (*e.g.* Tl solution for mass fractionation correction) could be added to the aqueous solution. Prohaska *et al.*[86] proposed an interface in which the end of the GC capillary was introduced into the sample inlet tube of a conventional Meinhard nebuliser which was connected to the end of the torch by a PTFE tube. [129]Xe was used as an internal standard. Even if condensation of the separated species occurred, the Meinhard nebuliser lead to the formation of an aerosol, which was transferred to the ICP without being adsorbed on the surface of the tube. Therefore no additional heating of the tube was required. An example of an interface design *via* a spray chamber is shown in Figure 3.8.

The disadvantage of systems that include a spray chamber is the relatively low sensitivity owing to the loss of energy in the desolvation of wet aerosol. Also, the applicability of this type of interface to less volatile species (Kovats indices above 5000) needs to be verified. The design fails with complex matrices which may condense in the spray chamber.[87]

5 Choice of the Mass Spectrometer

Quadrupole mass spectrometers (QMS) have predominantly been used, their sensitivity having improved a factor of ten during the last decade. Tao *et al.*[81] reported an instrumental detection limit of 0.7 fg for tin by operating the shield torch at normal plasma conditions using an HP 4500 instrument.

The ability to produce complete mass spectra at a high frequency makes TOF MS nearly ideal for the detection of transient signals produced by high speed chromatographic techniques.[10–12,88] In practice, however, no really significant advantage of ICP TOF MS over ICP Q MS has been demonstrated. The reason may be the absence of the need for the simultaneous measurement of more than three isotopes at a time and lower sensitivity in the monoelemental mode in comparison with the last generation of ICP quadrupole mass spectrometers. A minimum of approximately 0.5 ng of each species was found to be necessary for the measurement of isotope ratios with a precision better than 0.5%,[12] a quantity 100 times larger than those usually measured. On the other hand, Heisterkamp *et al.*[11] reported a detection limit (DL) of 10–15 fg for alkyllead compounds, a value which is comparable with those obtained by ICP Q MS.

A high resolution instrument may be required when the detection of sulfur or phosphorous is needed. Sanz-Medel's group proposed a technique for the detection of [32]S in GC analysis of volatile sulfur compounds in bad breath and reported, when a guard electrode was used under cold plasma conditions, absolute DLs in the low nanogram range.[89] Studies of natural biogeochemical isotopic fractionation correlated with a particular species will require the use of multi-collector detectors, either with a doubly focusing instrument[14] or with a single magnetic sector instrument equipped with a hexapole collision cell.[13]

6 GC-ICP MS Studies using Stable Isotopes

The isotopic specificity of ICP MS opens the way for the use of stable isotopes or stable isotope enriched species for studies of transformations and of artefact formation during sample preparation and to the wider implementation of isotope dilution quantification. The latter had until recently been limited by the non-availability of organometallic species with the isotopically enriched element. However, standards for isotopically enriched $Me^{201}Hg$,[90] monobutylin (MBT), dibutyltin (DBT), tributyltin (TBT)[91] and Me_3PbCl[92] have been recently synthesised and applications are being developed. The prerequisite of the use of isotope dilution techniques is the precise and accurate measurement of the isotopic ratios. To date applications to real-life samples have been exclusively carried out with ICP Q MS but precision and accuracy values for the measurement of isotope ratios in standard compounds by ICP TOF MS[10] and by sector-field multicollector instruments[13,14] have been reported. Tracer experiments can be performed with ICP Q MS but for natural fractionation studies the measurement precision may not be sufficient.

Precision and Accuracy of Isotope Ratio Measurements in GC-ICP MS

A precision of 1% has been reported for the Hg isotope ratios determined for MeEtHg eluted from a packed column by GC-ICP Q MS.[93] In capillary GC Q ICP MS a precision of 0.5% has been reported for Se (derivatised as piazselenol).[94] A tin isotope ratio measurement accuracy of 0.28% and a precision of 2.88% have been calculated for a 1 s wide GC peak of Me_4Sn.[10] Haas *et al.* reported that a minimum of 0.5 ng of an organometallic species was necessary for the measurement of isotope ratios with a precision better than 0.5%, the best value (0.34%) was attained for Me_2SnH_2.[12] ICP TOF MS gave 10 times better precision than ICP Q MS in comparable conditions when natural gas samples were analysed.[12]

The precision values reported for the measurement of the major Pb isotope ratio with a double-focusing ICP multicollector MS were below 0.07% (for a 3 s transient signal) and corresponded to an accuracy of 0.35%.[14] When a single magnetic sector instrument (with a hexapole collision cell and multicollector detection) was used the precision was in the range 0.02–0.07% for the ratio of the high-abundance isotopes and injections of 5–50 pg.[13] After mass bias correction the accuracy was 0.02–0.15%.[13]

For accurate determinations by the isotope dilution technique the mass discrimination effect (*ca.* 0.5% per mass unit) must be taken into account. Ways to measure and correct for the mass bias include the sequential measurement of the isotope ratios in the sample and the standard[95] or the addition of an internal standard, such as Cd[11,12] or Tl[13,14] and the simultaneous measurement of the $^{111}Cd/^{113}Cd$ or $^{203}Tl/^{205}Tl$ isotopic ratios, respectively. From a practical point of view the latter system requires the simultaneous delivery of the analyte and the internal standard to the ICP MS which can be done only *via* a spray chamber interface in view of the involatility of Cd and Tl species. An alternative measurement of the isotope ratio of the analyte element in a standard has been possible using to a diffusion cell containing a pure compound of the element to be determined.[95] It consists of a glass vial covered by a membrane which allows diffusion of the volatile calibrant species into the flow cell. If the isotope ratios of the element in the calibration compound are known, the measured isotope ratio of the separated species in the sample can be corrected.[95]

Monitoring of Artefacts During Sample Preparation Procedure

Enriched isotopes and isotope-enriched species provide an important diagnostic tool for the validation development of new analytical methods. The formation of MeHg$^+$ during a water vapour distillation procedure detected using isotopically enriched Hg^{2+}[93,96] has shed a new light on the accuracy of methylmercury determinations in sediment samples.[97] Stable isotope labelled mercury species have allowed a GC-ICP MS study of the simultaneous Hg^{2+} biomethylation and CH$_3$Hg$^+$ demethylation at ambient trace levels with a sensitivity superior to radiotracer techniques.[98]

Using a similar approach and a Me^{201}Hg spike, the transformation of MeHg$^+$ into elemental mercury (Hg^0) in the presence of chloride and bromide during derivatisation using NaBEt$_4$ but not by NaBPr$_4$ was demonstrated.[90] Isotope enriched Hg^{2+} allowed the observation of artifactual formation of methylmercury when water vapour distillation was applied to aqueous rain samples containing visible particles.[99] No artefact formation of methylmercury during sample preparation was observed following the addition of a ^{201}Hg^{2+} isotope standard.[100]

Speciated Isotope Dilution Analysis

Fundamentals of isotope dilution (ID) GC-ICP MS for species specific analysis have been extensively discussed by Gallus and Heuman.[95] Speciated ID analysis is only possible for element species well defined in their structure and composition. The species must not undergo interconversion and isotope exchange prior to separation. The equilibration of the spike and analyte, attainable in classical ID MS by multiple sequential dissolution and evaporation-to-dryness cycles, cannot be guaranteed to be achieved for speciated ID analysis in solid samples. Consequently, the prerequisite of the ID method, that the spike is added in an identical form to the analyte, is extremely difficult, if not impossible, to attain. Nevertheless, some benefits, such as the inherent corrections for the loss of analyte during sample preparation, for the incomplete derivatisation yield and for the intensity suppression/enhancement in the plasma are evident. In particular, ID quantification seems to be attractive in speciation analysis of complex matrices when the different organic consituents of the sample continuously modify the conditions in the plasma and thus, the sensitivity.[101]

Isotopically enriched species represent the ultimate means for specific, accurate and precise instrumental calibration. Not only are they useful for routine determination by reducing the analysis time, but they also assist in the testing and diagnostics of new analytical methods and techniques. To date, examples of the applications of speciated ID GC-ICP MS have been relatively scarce. The determination of dibutyltin in sediment has been carried out by ID analysis using a ^{118}Sn enriched spike. No recovery corrections for aqueous ethylation or extraction into hexane were necessary and no rearrangement reactions were evident from the isotope ratios.[102] A mixed spike containing ^{119}Sn enriched mono-, di- and tributyltin has been prepared by direct butylation of ^{119}Sn metal and characterised by reversed isotope dilution analysis by means of natural mono-, di- and tributyltin standards. A spike characterised in this way has been used for the simultaneous determination of the three butyltin compounds in sediment certified reference materials (CRMs).[91] Isotopically labelled Me_2Hg, $MeHgCl$ and $HgCl_2$ species have been prepared and used for the determination of the relevant species in gas condensates with detection limits in the low pg range.[101]

References

1. J.C. Van Loon, L.R. Alcock, W.H. Pinchin and J.B. French, *Spectrosc. Lett.*, 1986, **19**, 1125.
2. N.S. Chong and R.S. Houk, *Appl. Spectrosc.*, 1987, **41**, 66.
3. P.C. Uden, *J. Chromatogr.*, 1995, **703**, 393.
4. K. Sutton, R.M.C. Sutton and J.A. Caruso, *J. Chromatogr.*, 1997, **789**, 85.
5. R. Lobinski and F.C. Adams, *Spectrochim. Acta*, 1997, **52B**, 1865.
6. G.K. Zoorob, J.W. McKiernan and J.A. Caruso, *Mikrochim. Acta.*, 1998, **128**, 145.
7. B. Bouyssiere, J. Szpunar and R. Lobinski, *Spectrochim. Acta*, 2002, **57B**, 805.
8. B.W. Pack, J.A.C. Broekaert, J.P. Guzowski, J. Poehlman and G.M. Hieftje, *Anal. Chem.*, 1998, **70**, 3957.
9. F. Vanhaecke and L. Moens, *Fresenius' J. Anal. Chem.*, 1999, **364**, 440.
10. A.M. Leach, M. Heisterkamp, F.C. Adams and G.M. Hieftje, *J. Anal. At. Spectrom.*, 2000, **15**, 151.
11. M. Heisterkamp and F.C. Adams, *Fresenius' J. Anal. Chem.*, 2001, **370**, 597.
12. K. Hass, J. Feldmann, R. Wennrich and H.J. Stärk, *Fresenius J. Anal. Chem.*, 2001, **370**, 587.
13. E.M. Krupp, C. Pécheyran, S. Meffan-Main and O.F.X. Donard, *Fresenius J. Anal. Chem.*, 2001, **370**, 573.
14. E.M. Krupp, C. Pécheyran, H. Pinaly, M. Motelica-Heino, D. Koller, S.M.M. Young, I.B. Brenner and O.F.X. Donard, *Spectrochim. Acta*, 2001, **56B**, 1233.
15. R. Lobinski, I.R. Pereiro, H. Chassaigne, A. Wasik and J. Szpunar, *J. Anal. At. Spectrom.*, 1998, **13**, 859.
16. B. Bouyssiere, F. Baco, L. Savary and R. Lobinski, *Oil Gas Sci. Technol.*, 2000, **55**, 639.
17. W.P. Liu and H.K. Lee, *J. Chromatogr. A*, 1999, **834**, 45.

18. K. Jin, Y. Shibata and M. Morita, *Anal. Chem.*, 1991, **63**, 986.
19. J. Dedina and D.L. Tsalev, *Hydride-Generation Atomic-Absorption Spectrometry*, Wiley, Chichester, 1995.
20. F. Pannier, A. Astruc and M. Astruc, *Anal. Chim. Acta.*, 1996, **327**, 287.
21. E. Segovia Garcia, J.I. Garcia Alonso and A. Sanz-Medel, *J. Mass Spectrom.*, 1997, **32**, 542.
22. U.M. Grüter, M. Hitzke, J. Kresimon and A.V. Hirner, *J. Chromatogr. A*, 2001, **938**, 225.
23. I. Koch, J. Feldmann, J. Lintschinger, S.V. Serves, W.R. Cullen and K.J. Reimer, *Appl. Organomet. Chem.*, 1998, **12**, 129.
24. S. Rapsomanikis, *Analyst*, 1994, **119**, 1429.
25. M. Ceulemans and F.C. Adams, *J. Anal. At. Spectrom.*, 1996, **11**, 201.
26. M.B. de la Calle Guntinas, R. Lobinski and F.C. Adams, *J. Anal. At. Spectrom.*, 1995, **10**, 111.
27. T. De Smaele, L. Moens, R. Dams, P. Sandra, J. Van der Eycken and J. Vandyck, *J. Chromatogr. A*, 1998, **793**, 99.
28. K. Bergmann and B. Neidhart, *Fresenius' J. Anal. Chem.,* 1996, **356**, 57.
29. V. Minganti, R. Capelli and R. De Pellegrini, *Fresenius' J. Anal. Chem.*, 1995, **351**, 471.
30. R. Lobinski and F.C. Adams, *Anal. Chim. Acta*, 1992, **262**, 285.
31. R. Lobinski, J. Szpunar Lobinska, F.C. Adams, P.L. Teissedre and J.C. Cabanis, *J. Associ. Off. Anal. Chem.*, 1993, **76**, 1262.
32. R. Lobinski, C.F. Boutron, J.P. Candelone, S. Hong, J. Szpunar Lobinska and F.C. Adams, *Anal. Chem.*, 1993, **65**, 2510.
33. R. Lobinski, W.M.R. Dirkx, J. Szpunar-Lobinska and F.C. Adams, *Anal. Chim. Acta*, 1994, **286**, 381.
34. R. Lobinski, W.M.R. Dirkx, M. Ceulemans and F.C. Adams, *Anal. Chem.*, 1992, **64**, 159.
35. E. Bulska, D.C. Baxter and W. Frech, *Anal. Chim. Acta*, 1991, **12**, 545.
36. E. Bulska, H. Emteborg, D.C. Baxter, W. Frech, D. Ellingsen and Y. Thomassen, *Analyst*, 1992, **117**, 657.
37. G.B. Jiang and Q.F. Zhou, *J. Chromatogr. A*, 2000, **886**, 197.
38. G. Schwedt, *Chromatographic methods in inorganic analysis*, Hüthig, Heidelberg, 1981.
39. H. Kataoka, Y. Miyanaga and M. Makita, *J. Chromatogr.*, 1994, **659**, 481.
40. X.J. Cai, E. Block, P.C. Uden, X. Zhang, B.D. Quimby and J.J. Sullivan, *J. Agric. Food Chem.*, 1995, **43**, 1754.
41. K. Yasumoto, T. Suzuki and M. Yoshida, *J. Agric. Food. Chem.*, 1988, **36**, 463.
42. Z. Ouyang, J. Wu and L. Xie, *Anal. Biochem.*, 1989, **178**, 77.
43. O. Zheng and J. Wu, *Biomed. Chromatogr.*, 1988, **2**, 258.
44. R. Scerbo, M.B. de la Calle Guntinas, C. Brunori and R. Morabito, *Quim. Anal.*, 1997, 87.
45. F.A. Claussen, *J. Chromatogr. Sci.*, 1997, **35**, 568.
46. K. Schoene, J. Steinhanses, H.J. Bruckert and A. Koenig, *J. Chromatogr. A*, 1992, **605**, 257.

47. D. Amouroux, E. Tessier, C. Pecheyran and O.F.X. Donard, *Anal. Chim. Acta*, 1998, **377**, 241.
48. C. Pecheyran, C.R. Quetel, F.M. Martin and O.F.X. Donard, *Anal. Chem.*, 1998, **70**, 2639.
49. W. Dirkx, R. Lobinski and F.C. Adams, *Anal. Sci.*, 1993, **9**, 273.
50. R. Lobinski, V. Sidelnikov, Y. Patrushev, I. Rodriguez and A. Wasik, *Trends Anal. Chem.*, 1999, **18**, 449.
51. S. Slaets, F. Adams, I.R. Pereiro and R. Lobinski, *J. Anal. At. Spectrom*, 1999, **14**, 851.
52. I. Rodriguez, S. Mounicou, R. Lobinski, V. Sidelnikov, Y. Patrushev and M. Yamanaka, *Anal. Chem.*, 1999, **71**, 4534.
53. I. Rodriguez Pereiro, V.O. Schmitt and R. Lobinski, *Anal. Chem.*, 1997, **69**, 4799.
54. S. Perez Mendez, M. Montes Bayon, E. Blanco Gonzalez and A. Sanz-Medel, *J. Anal. At. Spectrom.*, 1999, **14**, 1333.
55. S. Rapsomanikis, O.F.X. Donard and J.H. Weber, *Anal. Chem.*, 1986, **58**, 35.
56. A. Wasik, I. Rodriguez Pereiro, C. Dietz, J. Szpunar and R. Lobinski, *Anal. Commun.*, 1998, **35**, 331.
57. A. Wasik, R. Lobinski and J. Namiesnik, *Instr. Sci. Technol.*, 2001, **29**, 393.
58. A. Wasik, I. Rodriguez Pereiro and R. Lobinski, *Spectrochim. Acta*, 1998, **53B**, 867.
59. Z. Mester and R. Sturgeon, *Spectrochim. Acta*, 2001, **56B**, 233.
60. L. Moens, T. De Smaele, R. Dams, P. Van Den Broek and P. Sandra, *Anal. Chem.*, 1997, **69**, 1604.
61. J. Vercauteren, A. De Meester, T. De Smaele, F. Vanhaecke, L. Moens, R. Dams and P. Sandra, *J. Anal. At. Spectrom.*, 2000, **15**, 651.
62. S. Aguerre, G. Lespes, V. Desauziers and M. Potin-Gautier, *J. Anal. At. Spectrom.*, 2001, **16**, 263.
63. S. Aguerre, C. Bancon-Montigny, G. Lespes and M. Potin-Gautier, *Analyst*, 2000, **125**, 263.
64. E. Baltussen, P. Sandra, F. David and C. Cramers, *J. Microcolumn Sep.*, 1999, **11**, 737.
65. J. Vercauteren, C. Pérès, C. Devos, P. Sandra, F. Vanhaecke and L. Moens, *Anal. Chem.*, 2001, **73**, 1509.
66. R. Lobinski and F.C. Adams, *J. Anal. At. Spectrom.*, 1992, **7**, 987.
67. C.M. Tseng, D. Amouroux, I.D. Brindle and O.F.X. Donard, *J. Environ. Monitor.*, 2000, **2**, 603.
68. E. Segovia Garcia, J.I. Garcia Alonso and A. Sanz-Medel, *J. Mass Spectrom.*, 1997, **32**, 542.
69. A.W. Kim, M.E. Foulkes, L. Ebdon, S.J. Hill, R.L. Patience, A.G. Barwise and S.J. Rowland, *J. Anal. At. Spectrom.*, 1992, **7**, 1147.
70. A. Kim, S. Hill, L. Ebdon and S. Rowland, *J. High. Res. Chromatogr.*, 1992, **15**, 665.
71. L. Ebdon, E.H. Evans, W.G. Pretorius and S.J. Rowland, *J. Anal. At. Spectrom.*, 1994, **9**, 939.

72. H.E.L. Armstrong, W.T. Corns, P.B. Stockwell, G. O'Connor, L. Ebdon and E.H. Evans, *Anal. Chim. Acta*, 1999, **390**, 245.
73. T. De Smaele, P. Verrept, L. Moens and R. Dams, *Spectrochim. Acta,* 1995, **50B**, 1409.
74. M. Montes Bayon, M. Gutierrez Camblor, J.I. Garcia Alonso and A. Sanz-Medel, *J. Anal. At. Spectrom.*, 1999, **14**, 1317.
75. I.A. Leal-Granadillo, J.I.G. Alonso and A. Sanz-Medel, *Anal. Chim. Acta*, 2000, **423**, 21.
76. W.G. Pretorius, L. Ebdon and S.J. Rowland, *J. Chromatogr.*, 1993, **646**, 369.
77. J. Poehlman, B.W. Pack and G.M. Hieftje, *Am. Lab.*, 1998, **30**.
78. B.W.P. Poehlman and G.M. Hieftje, *Int. Lab.*, 1999, **29**, 26.
79. C.R. Quétel, H. Tao, M. Tominaga and A. Miyazaki, *ICP Inf. Newsl.*, 1996, **21**, 77.
80. H. Tao, T. Murakami, M. Tominaga and A. Miyazaki, *J. Anal. At. Spectrom.*, 1998, **13**, 1085.
81. H. Tao, C.R. Ramaswamy Babu Rajendran, C.R. Quetel, T. Nakazato, M. Tominaga and A. Milyazaki, *Anal. Chem.*, 1999, **71**, 4208.
82. R.B. Rajendran, H. Tao, T. Nakazato and A. Miyazaki, *Analyst*, 2000, **125**, 1757.
83. G.R. Peters and D. Beauchemin, *J. Anal. At. Spectrom.*, 1992, **7**, 965.
84. G.R. Peters and D. Beauchemin, *Anal. Chem.*, 1993, **65**, 97.
85. J. Feldmann, R. Gruemping and A.V. Hirner, *Fresenius J. Anal. Chem.*, 1994, **350**, 228.
86. T. Prohaska, M. Pfeffer, M. Tulipan, G. Stingeder, A. Mentler and W.W. Wenzel, *Fresenius J. Anal. Chem.*, 1999, **364**, 467.
87. B. Bouyssiere, *unpublished*, personal communication.
88. J.R. Baena, M. Gallego, M. Valcarcel, J. Leenaers and F.C. Adams, *Anal. Chem.*, 2001, **73**, 3927.
89. J. Rodriguez-Fernandez, M. Montes-Bayon, R. Pereiro and A. Sanz-Medel, *J. Anal. At. Spectrom.*, 2001, **16**, 1051.
90. N. Demuth and K.G. Heumann, *Anal. Chem.*, 2001, **73**, 4020.
91. J.R. Encinar, M.I.M. Villar, V.G. Santamaria, J.I.G. Alonso and A. Sanz-Medel, *Anal. Chem.*, 2001, **73**, 3174.
92. L. Ebdon, S.J. Hill and C. Rivas, *Spectrochim. Acta,* 1998, **53B**, 289.
93. H. Hintelmann, R.D. Evans and J.Y. Villeneuve, *J. Anal. At. Spectrom.*, 1995, **10**, 619.
94. K.G. Heumann, S.M. Gallus, G. Rädlinger and J. Vogl, *Spectrochim. Acta*, 1998, **53B**, 273.
95. S.M. Gallus and K.G. Heumann, *J. Anal. At. Spectrom.*, 1996, **11**, 887.
96. H. Hintelmann, R. Falter, G. Ilgen and R.D. Evans, *Fresenius' J. Anal. Chem.*, 1997, **358**, 363.
97. P. Quevauviller, F. Adams, J. Caruso, M. Coquery, R. Cornelis, O.F.X. Donard, L. Ebdon, M. Horvat, R. Lobinski, R. Marabito, H. Muntau and M. Valcarcel, *Anal. Chem*, 1999, **71**, 155A.
98. H. Hintelmann and R.D. Evans, *Fresenius' J. Anal. Chem.*, 1997, **358**, 378.

99. J. Holz, J. Kreutzmann, R.-D. Wilken and R. Falter, *Appl. Organomet. Chem.*, 1999, **13**, 789.
100. Q. Tu, J. Qian and W. Frech, *J. Anal. At. Spectrom.*, 2000, **15**, 1583.
101. J.P. Snell, I.I. Stewart, R.E. Sturgeon and W. Frech, *J. Anal. At. Spectrom.*, 2000, **15**, 1540.
102. J.R. Encinar, J.I. Garcia Alonso and A. Sanz-Medel, *J. Anal. At. Spectrom.*, 2000, **15**.

CHAPTER 4

Liquid Chromatography with ICP MS Detection

1 Introduction

Most elemental species of interest in speciation analysis are non-volatile and cannot be converted into such by means of derivatisation. They include virtually all the coordination complexes of trace metals but also many truly organometallic (containing a covalently bound metal or metalloid) compounds. For all these species HPLC is the principal separation technique prior to element selective detection. Alternatively capillary and gel electrophoresis (discussed in Chapter 5) can be used but their practical significance is limited to some particular applications.

The coupling of HPLC to ICP MS is technically very simple and consists of connecting the exit of the chromatographic column to the nebuliser. The optimisation of the interface is limited to the choice of the nebuliser matching the flow rate from the column and to assuring the stability of the plasma in the presence of the mobile phase.

In spite of the elevated acquisition and running costs ICP MS is practically the sole element selective detector used for the HPLC of real-life samples. The coupling of HPLC and ICP MS offers an unmatched performance for the detection and determination of involatile metallospecies in plant and animal biochemistry, nutrition and clinical chemistry. Owing to the number of separation mechanisms available in LC, HPLC-ICP MS is an efficient and versatile technique to monitor trace quantities of metal complexes. The potential of HPLC-ICP MS for trace metal speciation has been widely discussed.[1-5]

HPLC-ICP MS offers multielement capability, low detection limits and the possibility of on-line isotope dilution. Indeed, with the sub ng L^{-1} detection limits offered by the (most popular) quadrupole analysers, ICP MS allows the detection of HPLC signals from as little as $0.1 \, ng \, mL^{-1}$ of an element in the injected solution. HPLC-ICP MS with enriched stable isotopes is a unique analytical method by which speciation of both endogenous elements and external tracers can be achieved in a single experiment.[6]

HPLC-ICP MS is a valuable alternative technique for validating GC based techniques in speciation studies. In routine analysis, however, GC should be used whenever possible because of the higher separation efficiency, detection sensitivity and lesser vulnerability of the column to fouling with dirty matrices. This chapter focuses on the role of HPLC-ICP MS in speciation studies which cannot be carried out by any other hyphenated technique.

2 Separation of Elemental Species by Liquid Chromatography

The principal HPLC separation mechanisms used in metal speciation analysis include size exclusion, ion exchange and reversed phase chromatography. Because of the complex nature of metal-biomolecules systems, a combination of these separation mechanisms is often needed to identify correctly the trace element species. The choice of the separation mechanism depends on the physicochemical properties of the analyte species. The requirements and limitations of the interface between HPLC and ICP MS vary as a function of the flow-rate, composition of the mobile phase used and the analytical objectives to be achieved. Separations of trace (picogram) quantities of metal complexes with biomolecules are still under development and many problems, such as unspecific adsorption, species decomposition and unpredictable memory effects occur.

It should be noted that the coelution of a metal and a particular biomacromolecule is only an indication (not a proof!) that they belong together. The ultimate confirmation can be obtained by applying affinity chromatography for the particular protein. The presence of the metal attached to the protein bound to the antigen on the column confirms the existence of the metal-protein complex.[7]

In terms of sample preparation, the filtration of the cytosol using a 0.45 μm or, better, a 0.22 μm filter before injection onto the chromatographic column is mandatory. A guard column should be used to protect the analytical column particularly from the effects of lipids which would otherwise degrade the separation. A number of bioanalytical techniques including ultracentrifugation, microdialysis and ammonium sulfate precipitation often precede the chromatographic separations of metallocompounds. Successive ultrafiltration through membranes with molecular weight cut-offs of 30 000, 5000 and 500 u which has been used to study the distribution of metal species as a function of molecular weight[8-10] and is recommended for the simplification of the matrix loaded on a HPLC column.

Size Exclusion Chromatography

Size exclusion chromatography (SEC) is based on the molecular sieve effect and enables species to be separated according to their size and to a lesser extent, shape. The average time a substance spends in the pores (determined, for a given shape, by the size) can usually be correlated with its molecular weight. Figure 4.1a shows a calibration graph for a size exclusion column allowing the evaluation of the molecular weight of an eluting compound. A typical chromatogram is shown in Figure 4.1b.

The accuracy of the molecular mass evaluation is acceptable only for fairly large proteins and polysaccharides. For smaller species, especially ions with a high charge-to-mass ratio, secondary adsorption and ion exchange effects affect the separations. These phenomena, initially considered to be a nuisance in SEC are becoming employed more and more often for the separation of organoselenium[11,12] and organoarsenic compounds.[13–16]

SE HPLC has the advantage over other LC techniques because of the high tolerance to the matrix, the compatibility of the mobile phase flow rates used

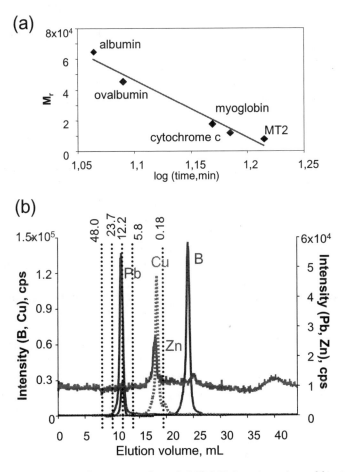

Figure 4.1 *Size exclusion chromatography with ICP MS detection: a) a calibration graph; b) a typical chromatogram (a red wine sample) allowing the estimation of the molecular mass of eluting metallocompounds*

(0.7–1.0 mL/min) and compositions with AAS and ICP spectrometers. Another potential advantage is the possibility of avoiding the buffer[11,17] and hence the possibility of a simplification of the matrix in heart-cut and lyophilised fractions. A limitation is the multifold dilution of the sample during chromatography.

Packing

Separation by SE HPLC should be independent of the charge of the analyte but in practice the stationary phase surface displays charged properties so that a mixed-mode separation may be observed. This makes the choice of packing critical. The categories of packing used include silica and organic polymers. At the nanogram levels involved significant silanophilic effects including metal losses in the presence of low ionic strength eluents on silica-based SEC supports have been reported.[18–21] A copolymeric styrene-divinylbenzene SEC support that provided a symmetrical peak with negligible losses of Cd during chromatography of Cd-metallothionein complexes has been proposed.[19,20] The average pore size of the packing used varies from 100 to 1000 Å. For detailed speciation in complex samples such as some body fluids, columns with a separation range up to 250 kDa should be used.[22] The combination of two SEC columns with different fractionation ranges may offer an attractive alternative.[22]

Mobile Phase

The optimum eluent should assure minimum competition between buffer and cytosolic ligands, and between these ligands and the gel. Carrying out the separation with water as the mobile phase has been reported to prevent structural changes, denaturation of proteins, and destruction of protein-metal complexes.[23] On the other hand, even dilute buffers generally, cause adsorption of low molecular weight proteins onto the column packing. Therefore, various aqueous mobile phases of fairly high ionic strength have been used to avoid interactions with the packing material. When silica-based packings cannot be avoided, the addition of a non-complexing salt (*e.g.* 0.1 M NaCl) to the mobile phase is necessary in order to suppress the residual silanol activity of the column packing.[21]

The addition of EDTA proposed by some authors[24] to minimise metal ion-gel interactions in chromatography of metal-protein complexes was found to be unsuitable by others[25] because of the occasional presence of anomalous Cd peaks in subsequent runs. Polymeric supports suffer from the deposition of excess free Cd^{2+} which interacts with the analytes, often causing severe degradation in peak resolution.[20] Because of the weak complexing character Tris buffer is not sufficient to compete with the polymeric support for Cd^{2+}; complexation with β-mercaptoethanol is advised.[20]

The wide variety of buffers reported in the literature makes it relatively easy to choose one suitable for the detection technique to be used. Up to 50 mM Tris-HCl was found to be well tolerated by ICP MS whereas 20 mM formate or acetate buffer in 10% methanol are acceptable for ES MS. The addition of 0.03% NaN_3 as a retardant of bacterial activity is advised to protect the column from damage resulting from bacterial growth when real-life samples are analysed.

Analysis Time

This is a function of the column size and the flow rate. Although columns up to 1200 mm are used the standard 300×7.6 mm column is a good choice. At a flow rate of 1 mL min^{-1}, the separation requires *ca.* 20 min to complete. The choice of small-bore columns with size exclusion packings is still limited[26] but such columns enable the rapid characterisation of various metal-containing molecular weight fractions in unknown cytosols by direct injection nebulisation ICP MS.

Separation Efficiency

The number of theoretical plates in SE HPLC is small; this technique is not only insufficient for the discrimination of the small amino acid heterogeneities in the chromatography of metallopeptides, but also lacks the resolution for frequently encountered problems, such as the separation of serum selenoproteins or the separation of human albumin and transferrin. Each fraction eluted from a size exclusion column may still contain hundreds of compounds. In most cases further signal characterisation by orthogonal (complementary) chromatographic techniques is necessary.

Ion Exchange Chromatography

Ion exchange chromatography is based on the interactions of analyte cations in the mobile phase with the negatively charged functional groups of the stationary phase (cation exchange (CE)) or of analyte anions with positively charged functional groups (anion exchange (AE)). In the high performance version (sometimes termed ion chromatography (IC)) polymer beads are coated with ion exchange groups. Both cation and anion exchange have been widely used for the separation of metal species, especially of organoarsenic and organoselenium compounds. In metalloprotein biochemistry the technique has been applied to the fractionation of metallothionein[27,28] and serum proteins.[29-31] Separations on ion exchange resins require eluents containing high salt concentrations in order to elute multicharged anions and cations in a reasonable time.

Weak anion exchangers with diethylaminoethyl functional groups have been used in the majority of work on the separation of metal complexes with metallothioneins.[32] A strong quaternary ammonium salt anion exchanger has been used to separate aluminium carrying serum proteins.[29] It allowed the separation of albumin and transferrin leaving the metals attached to these proteins. Strong anion exchangers were also preferred for the separation of arsenic and selenium species.

Aqueous buffers with a linear concentration gradient have been used as eluents. In the separation of proteins, buffers used in anion exchange chromatography

often exceed concentrations of 0.1 M. This is likely to result in instability of the
sensitivity of the ICP system in the long term because of clogging of the nebuliser
and sampler and skimmer cones. For example, the high potential of separation of
the MT-1 and MT-2 isoforms by anion exchange has not been fully exploited in
the coupled system because the common final concentration of 0.25 M of buffer
is difficult for the ICP MS to tolerate. Much lower buffer concentrations have
been used in AE chromatography of organometalloid species. Buffer concentrations of 3–20 mM seemed to be sufficient.[33–40]

Cation exchange HPLC has been less popular than anion exchange HPLC.
However, since the separations can be carried out in acidic media using several
mM pyridine-formate buffer[33,41] or simply mM HNO_3 eluents,[42,43] the chromatographic conditions developed can be directly transferred to HPLC-ES MS.[44,45]

Reversed Phase Chromatography

Analytes in a polar mobile phase, such as water or water-methanol, are chromatographed using a relatively non-polar stationary phase – silica gel containing
chemically (covalently) bonded hydrocarbon groups $(C_4–C_{18})$. Polar, uncharged
compounds with an $M_r < 3000$ u are typically analysed. However, the number of
applications for proteins is constantly increasing because of the excellent
resolution and ablity to differentiate proteins, such as metallothioneins varying by
only one amino acid.[32] The primary areas of application of reversed phase (RP)
HPLC in bioinorganic analytical chemistry include the separation of species in
10 kDa ultrafiltrates of samples or sample extracts and the accurate characterisation of purity of metallothionein fractions isolated by size exclusion and anion
exchange chromatography. The use of ion pairing reagents in the mobile phase
(ion interaction chromatography) allows the extension of RP HPLC to ionic
analytes. In particular, the latter mode is employed for speciation of organoarsenic
and organoselenium compounds. It should be noted that the use of ion pairing
reagents in the mobile phase dramatically reduces the ES MS detection
sensitivity.

Reversed phase HPLC seems to be superior to SE HPLC and IE HPLC for the
separation of metal-biomolecule complexes because the packing material for RP
HPLC is largely free of metals ligands.[46] Since the hydrophobicity of a
biomolecule primarily dictates its retention in RP HPLC, the gradual elution of
the individual species of a mixture is achieved by decreasing the polarity of the
mobile phase by the gradient addition of methanol or acetonitrile. The separation
of metal complexes with metallothioneins (MTs) by RP HPLC has been
reviewed.[32,47]

The availability of RP columns with different dimensions makes it possible to
adjust the sample introduction flow rate to that required by the detection system.
The flow rate depends strongly on the column geometry. Varying the column
inner diameter from 8 mm to 0.18 mm allows a change in the flow rate from
10 mL min^{-1} to 2 μL min^{-1} according to requirements. Standard bore (4–
5 mm i.d.) 15–25 cm long columns are the most frequently used. Narrow bore

and microbore columns are expected to gain in significance because of their higher sensitivity and resolution[48-50] but a microflow nebuliser is required for the introduction of the eluent into ICP MS. The use of 10–50 mM of buffer (usually ammonium acetate or Tris-HCl) is usually necessary. Microbore RP HPLC-ES MS has been optimised for the characterisation of metallothioneins.[51-56]

The high content of organic modifier makes RP HPLC poorly compatible with ICP MS. On the other hand, RP separation conditions are often close to ideal for ES MS detection.

3 Interface Between HPLC and ICP MS

Interfacing via a Concentric or Cross-Flow Nebuliser

The basic interface (Figure 4.2) is straightforward – a piece of narrow bore tubing that connects the outlet of the LC column with the liquid flow inlet of the nebuliser. Typical LC flow rates of 0.7–1.5 mL min^{-1} are within the range usually required for pneumatic nebulisation. The limitation is the low (1–5%) transfer efficiency, losses in the spray chamber and thus degraded sensitivity. Aqueous eluents with the buffer content up to 50 mM are tolerated. The transfer efficiency can be increased by using a hydraulic high pressure (HHPN)[33,57,58] or an ultrasonic nebuliser.[59-61]

Salts can cause short-term signal depression or enhancement and cause blockage of the nebuliser and the sampling cone. Concentrations of 50 mM phosphate buffers have been reported to cause the rapid erosion and clogging of the nickel sampling cone.[62]

Organic solvents used in RP HPLC negatively influence the stability of an ICP (until extinction in extreme cases). In particular, they are responsible for the deposition of carbon on the sampling cone and torch. The use of a water-cooled spray chamber and an increase in RF power can help to reduce the solvent load on the plasma and to increase its stability. A dedicated desolvation unit can cope with the methanol concentrations usually used in reversed-phase chromatography. The addition of oxygen (1–3%) to the nebuliser gas flow can help to minimise

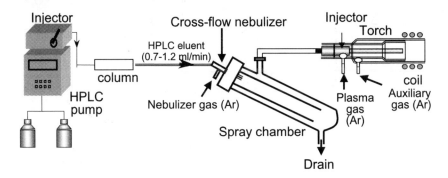

Figure 4.2 *The basic HPLC-ICP MS interface via a cross-flow nebuliser*

carbon deposition and clogging of the sample cone at the expense of its lifetime.[48,63-66] The use of Pt cones then becomes mandatory. Loss in sensitivity apparently cannot be avoided for higher concentrations of the organic modifier.

Interfacing of Microbore HPLC with ICP

In the case of microbore chromatography, the use of a direct injection nebuliser (DIN) interface appears to be particularly attractive. The DIN is a microconcentric

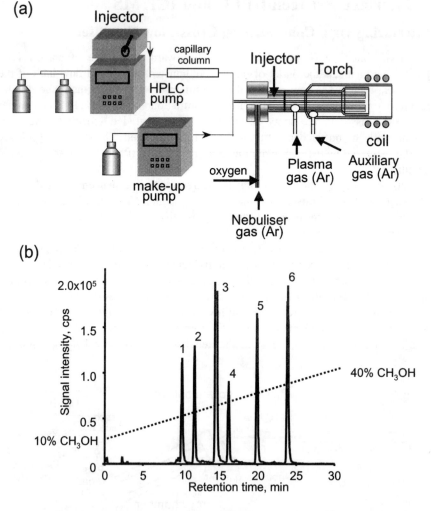

Figure 4.3 *Microbore reversed phase HPLC-ICP MS interface based on a direct injection nebuliser (DIN): a) interface design; b) chromatogram of a mixture of cobalamin derivatives*
(from ref. 65)

pneumatic nebuliser with no spray chamber; it nebulises the liquid sample directly into the central channel of the ICP torch. The low dead volume ($<2\ \mu L$) and the absence of a spray chamber in the DIN minimise post-column peak broadening and facilitate the use of low flow rates ($30-100\ \mu L\ min^{-1}$).[26,48,65,67] Another benefit

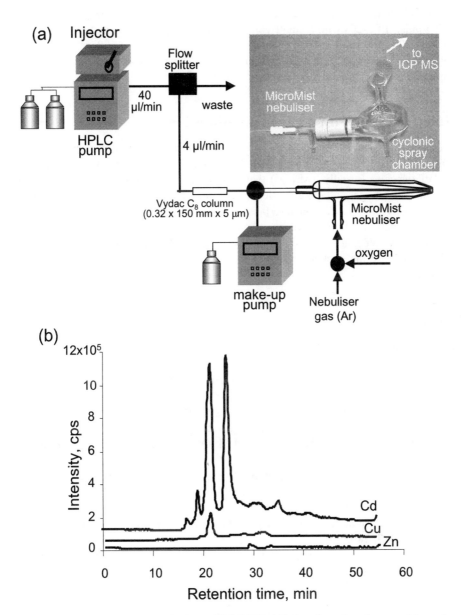

Figure 4.4 *Microbore reversed phase HPLC-ICP MS interface based on a Micromist nebuliser (Glass Expansions™): a) interface design; b) chromatogram of metallothionein isoforms obtained with a Micromist interface*

is the fast sample washout with minimal memory effects. Post-column makeup with water to lower the concentration of the organic modifier is a possible remedy, however, at the expense of sensitivity.[68] The interface design and a typical chromatogram for the separation of a mixture of cobalamin analogues are shown in Figure 4.3.

An alternative is the use of a microconcentric nebuliser (*e.g.* MCN-100, Micromist, PFE-20) with a small volume cyclonic spray chamber. A makeup flow of *ca.* 100–300 μL min^{-1} is necessary to match the optimum flow rate of the nebuliser. The interface design and an example chromatogram for the separation of metallothionein isoforms are shown in Figure 4.4.

Interfacing via Post-Column Volatilisation

The need for a nebuliser between an HPLC column and an ICP MS spectrometer can be eliminated by the post-column conversion of the elemental species into volatile species (usually hydrides) that can be swept into the plasma (*cf.* Chapter 2). In comparison with pneumatic nebulisation the on-line microwave assisted digestion-hydride generation interface offers 20–100-fold increase in sensitivity and elimination of interferences from the sample matrix or mobile phase components. When a sensitive detector, such as ICP MS, is used the reagent blank values increase the background noise and thus reduce the gain in detection limit to a factor of 2–10.[69]

References

1. G.K. Zoorob, J.W. McKiernan and J.A. Caruso, *Mikrochim. Acta*, 1998, **128**, 145.
2. K.L. Sutton, R.M.C. Sutton and J.A. Caruso, *J. Chromatogr. A*, 1997, **789**, 85.
3. J. Szpunar, *Analyst*, 2000, **125**, 963 .
4. B. Michalke, *Trends Anal. Chem.*, 2002, **21**, 154.
5. B. Michalke, *Trends Anal. Chem.*, 2002, **21**, 142.
6. K.T. Suzuki, S. Yoneda, M. Itoh and M. Ohmichi, *J. Chromatogr. B, Biomed. Appl.*, 1995, **670**, 63.
7. R. Cornelis, J. De Kimpe and X. Zhang, *Spectrochim. Acta, Part B*, 1998, **53B**, 187.
8. K. Lange Hesse, *Fresenius' J. Anal. Chem.*, 1994, **350**, 68.
9. K. Lange Hesse, L. Dunemann and G. Schwedt, *Fresenius' J. Anal. Chem.*, 1991, **339**, 240.
10. K. Lange Hesse, L. Dunemann and G. Schwedt, *Fresenius' J. Anal. Chem.*, 1994, **349**, 460.
11. C. Casiot, J. Szpunar, R. Lobinski and M. Potin Gautier, *J. Anal. At. Spectrom.*, 1999, **14**, 645.
12. C. Casiot, V. Vacchina, H. Chassaigne, J. Szpunar, M. Potin-Gautier and R. Lobinski, *Anal. Commun.*, 1999, **36**, 77.
13. M. Morita and Y. Shibata, *Anal. Sci.*, 1987, **3**, 575.

14. Y. Shibata and M. Morita, *Anal. Sci.*, 1989, **5**, 107.
15. Y. Shibata and M. Morita, *Anal. Chem.*, 1989, **61**, 2116.
16. J. Yoshinaga, Y. Shibata, T. Horiguchi and M. Morita, *Accred. Qual. Assur.*, 1997, **2**, 154.
17. B. Michalke, D.C. Muench and P. Schramel, *Fresenius' J. Anal. Chem.*, 1992, **344**, 306.
18. D. Klueppel, N. Jakubowski, J. Messerschmidt, D. Stuewer and D. Klockow, *J. Anal. At. Spectrom.*, 1998, **13**, 255.
19. K.A. High, J.S. Blais, B.A.J. Methven and J.W. McLaren, *Analyst*, 1995, **120**, 629.
20. K.A. High, B.A. Methven, J.W. McLaren, K.W.M. Siu, J. Wang, J.F. Klaverkamp and J.S. Blais, *Fresenius' J. Anal. Chem.*, 1995, **351**, 393.
21. S. Micallef, Y. Couillard, P.G.C. Campbell and A. Tessier, *Talanta*, 1992, **39**, 1073.
22. Y. Makino and S. Nishimura, *J. Chromatogr. B*, 1992, **117**, 346.
23. B. Michalke and P. Schramel, *J. Trace Elem. Electrolytes Health Dis.*, 1990, **4**, 163.
24. K. Takatera and T. Watanabe, *Anal. Sci.*, 1992, **8**, 469.
25. H.M. Crews, J.R. Dean, L. Ebdon and R.C. Massey, *Analyst*, 1989, **114**, 895.
26. S.C.K. Shum and R.S. Houk, *Anal. Chem.*, 1993, **65**, 2972.
27. A.H. Pan, F. Tie, B.G. Ru, L.Y. Li and T. Shen, *Biomed. Chromatogr.*, 1992, **6**, 205.
28. L.D. Lehman and C.D. Klaassen, *Anal. Biochem.*, 1986, **153**, 305.
29. A.B. Soldado Cabezuelo, E. Blanco Gonzalez and A. Sanz-Medel, *Analyst*, 1997, **122**, 573.
30. K. Wrobel, E. Blanco Gonzalez, K. Wrobel and A. Sanz-Medel, *Analyst*, 1995, **120**, 809.
31. A.B. Solado Cabezuelo, M. Montes Bayon, E. Blanco Gonzalez, J.I. Garcia Alonso and A. Sanz-Medel, *Analyst*, 1998, **123**, 865.
32. R. Lobinski, H. Chassaigne and J. Szpunar, *Talanta*, 1998, **46**, 271.
33. W. Goessler, W. Maher, K.J. Irgolic, D. Kuehnelt, C. Schlagenhaufen and T. Kaise, *Fresenius' J. Anal. Chem.*, 1997, **359**, 434.
34. W. Goessler, A. Rudorfer, E.A. Mackey, P.R. Becker and K.J. Irgolic, *Appl. Organometal. Chem.*, 1998, **12**, 491.
35. D. Velez, N. Ybanez and R. Montoro, *J. Anal. At. Spectrom.*, 1997, **12**, 91.
36. N. Ybanez, D. Velez, W. Tejedor and R. Montoro, *J. Anal. At. Spectrom.*, 1995, **10**, 459.
37. Z. Slejkovec, J.T. Van Elteren and A.R. Byrne, *Talanta*, 1999, **49**, 619.
38. H.M. Crews, P.A. Clarke, D.J. Lewis, L.M. Owen, P.R. Strutt and A. Izquierdo, *J. Anal. At. Spectrom.*, 1996, **11**, 1177.
39. S.M. Bird, H. Ge, P.C. Uden, J.F. Tyson, E. Block and E. Denoyer, *J. Chromatogr. A*, 1997, **789**, 349.
40. G. Alsing Pedersen and E.H. Larsen, *Fresenius' J. Anal. Chem.*, 1997, **358**, 591.
41. W. Goessler, D. Kuehnelt, C. Schlagenhaufen, Z. Slejkovec and K.J. Irgolic, *J. Anal. At. Spectrom.*, 1998, **13**, 183.

42. Y. Inoue, Y. Date, K. Yoshida, H. Chen and G. Endo, *Appl. Organometal. Chem.*, 1996, **10**, 707 .
43. Y. Inoue, Y. Date, T. Sakai, N. Shimizu, K. Yoshida, H. Chen, K. Kuroda and G. Endo, *Appl. Organometal. Chem.*, 1999, **13**, 81 .
44. J.J. Corr and E.H. Larsen, *J. Anal. At. Spectrom.*, 1996, **11**, 1215.
45. J.J. Corr, *J. Anal. At. Spectrom.*, 1997, **12**, 537.
46. H. Van Beek and A.J. Baars, *J. Chromatogr. A*, 1988, **442**, 345.
47. M.P. Richards, *Meth. Enzym.*, 1991, **205**, 217.
48. H. Chassaigne and R. Lobinski, *Anal. Chim. Acta.*, 1998, **359**, 227.
49. S.A. Pergantis, W. Winnik and D. Betowski, *J. Anal. At. Spectrom.*, 1997, **12**, 531.
50. S.A. Pergantis, E.M. Heithmar and T.A. Hinners, *Analyst*, 1997, **122**, 1063.
51. H. Chassaigne and R. Lobinski, *Anal. Chem.*, 1998, **70**, 2536.
52. H. Chassaigne and R. Lobinski, *Analyst*, 1998, **123**, 2125.
53. H. Chassaigne and R. Lobinski, *J. Chromatogr. A*, 1998, **829**, 127.
54. H. Chassaigne and R. Lobinski, *Talanta*, 1999, **48**, 109.
55. H. Chassaigne and R. Lobinski, *Fresenius' J. Anal. Chem.*, 1998, **361**, 267.
56. J.C.Y. Le Blanc, *J. Anal. At. Spectrom.*, 1997, **12**, 525.
57. N. Jakubowski, C. Thomas, D. Klueppel and D. Stuewer, *Analysis*, 1998, **26**, M37.
58. J. Zheng, W. Goessler and W. Kosmus, *Mikrochim. Acta*, 1998, **130**, 71.
59. S. Lustig, B. Michalke, W. Beck and P. Schramel, *Fresenius' J. Anal. Chem.*, 1998, **360**, 18.
60. A. Mazzucotelli, V. Bavastello, E. Magi, P. Rivaro and C. Tomba, *Anal. Proc*, 1995, **32**, 165.
61. K.L. Yang and S.J. Jiang, *Anal. Chim. Acta*, 1995, **307**, 109.
62. D. Heitkemper, J. Creed, J. Caruso and F.L. Fricke, *J. Anal. At. Spectrom.*, 1989, **4**, 279.
63. K. Takatera, N. Osaki, H. Yamaguchi and T. Watanabe, *Anal. Sci.*, 1994, **10**, 567.
64. K.L. Oedegard, W. Lund, *J. Anal. At. Spectrom.*, 1997, **12**, 403.
65. H. Chassaigne and J. Szpunar, *Analusis*, 1998, **26**, M48.
66. A. Makarov and J. Szpunar, *J. Anal. At. Spectrom.*, 1999, **14**, 1323.
67. H. Emteborg, G. Bordin and A.R. Rodriguez, *Analyst*, 1998, **123**, 245.
68. R. Lobinski, I.R. Pereiro, H. Chassaigne, A. Wasik and J. Szpunar, *J. Anal. At. Spectrom.*, 1998, **13**, 859.
69. J.M. Gonzalez LaFuente, J.M. Marchante-Gayon, M.L. Fernandez Sanchez, A. Sanz-Medel, *Talanta*, 1999, **50**, 207.

CHAPTER 5

Electrophoretic Techniques with Element Selective Detection

1 Introduction

The complexity of many environmental and biological matrices and the need to discriminate among analytes that have very close physicochemical properties (*e.g.* metalloproteins that vary by one amino acid only) prompted the development of high-resolution separation techniques prior to sensitive element- or molecule-specific detection.[1-3] These techniques are based on separations in an electric field. They include different mechanisms of capillary electrophoresis and flat-bed electrophoresis.

Capillary zone electrophoresis has a proven potential for the separation of redox states, small non-volatile metal complexes and metallopeptides. Detection limits down to $1\ \mu g\,L^{-1}$ can be obtained by the CZE-ICP MS coupling using the latest generation of quadrupole mass spectrometers.

Two-dimensional electrophoresis with its impressive resolution is a well established tool in proteomics and is being intensely developed for selenoproteomics. Applications for coordination complexes are limited by the problem of insufficient stability of the metal-protein species on the gel. For detection, laser ablation ICP MS is an elegant tool but improvements in sensitivity are still required.

The use of electrophoretic techniques for speciation research is relatively new and many problems are still present.[4] This chapter discusses the state-of-the-art in speciation analysis by electrophoresis with ICP MS detection.

2 Flatbed Gel Electrophoresis

Separation of Metal Species by Flatbed Electrophoresis

Flat-bed electrophoresis with its various formats such as polyacrylamide gel electrophoresis (PAGE), isoelectric focusing (IEF) and immunoelectrophoresis offers a number of attractive features for the characterisation of metallobiomolecules owing to its micropreparative loading and high resolution. Thousands of proteins can be separated in a single run on a two-dimensional system.

PAGE is based on converting all proteins into similar structures which differ only in terms of molecular mass. It is achieved by attaching a number of sodium dodecyl sulfate (SDS) molecules to a protein which results in its denaturation (breaking the 3-D tertiary structure). As a result all proteins regardless of their identity are imparted the same free-solution mobility so their separation is controlled by the molecular mass. Proteins with small m/z move fastest and farthest, proteins with large m/z move slowest. The standard SDS-PAGE protocols lead to high resolution and reproducibility but include chemicals such as SDS and dithiotreithol which denature the protein structure and cleave protein-metal bonds. Native techniques keep the proteins in their original conformation, but result in poorer resolution. Concentrations as low as 0.1% SDS have been applied to the purification of proteins under non-denaturating conditions.[5,6]

SDS-PAGE may not assure the complete dissociation of multimeric proteins into their subunits and thereby give rise to several labelled bands originating from the same compound. Tissues may contain proteins with similar migration velocities and an orthogonal separation mechanism is necessary. This is achieved by two-dimensional (2-D) electrophoresis.

Proteins are separated in the gradient gel according to the mass-to-charge ratio m/z in the first dimension. A second dimension can be added by isoelectric focusing using immobilised pH gradients. Proteins are separated by their isoelectric points in the immobilised pH gradient of the IEF strip gel. The charged proteins migrate within the pH gradient until they reach the pH in the gradient equal to their pI value. At that point their charge is zero and they focus as distinct zones. 2-D electrophoresis is widely used for the characterisation of proteins and biomolecules but its application to metal speciation has been scarce.[5,6] The principles of two-dimensional gel separations by IEF and PAGE are schematically shown in Figure 5.1.

The second dimension can also be achieved by immunoelectrophoresis in which the second-dimensional gel contains antibodies against the sample (human serum) proteins.[7,8]

Element Specific Detection in Gel Electrophoresis

The amount of proteins concentrated in the tiny gel volumes is very small and is thus hardly accessible to standard analytical chemical methods. Metal detection by autoradiography is the most elegant way at the experimental stage but requires the analyte species to contain a radioactive isotope. An emerging alternative is laser ablation ICP MS which offers quasi on-line resolution.[8,9]

With regard to speciation analysis, autoradiography has been extensively applied to the detection of selenoproteins separated by gel electrophoresis.[10-15] For example, the ^{75}Se nuclide ($t_{1/2} = 120.4$ days and emits γ-rays in the range 10 to 280 keV) is well suited for such experiments. Selenoproteins in an extract of the tissue homogenate of organisms labelled *in vivo* by administering ^{75}Se have been

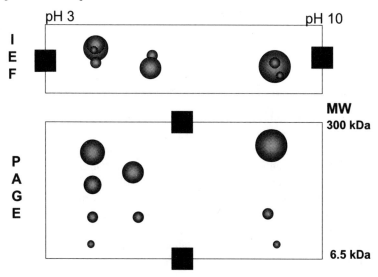

Figure 5.1 *Principles of 2-D flat-bed electrophoresis: IEF first dimension; PAGE second dimension*

separated by SDS-PAGE and the distribution of the tracer on the gel determined by autoradiography.[10–15]

The advantage of autoradiography is the very low limit of detection (in the sub-picogram or even femtogram range) which may be controlled by the uptake and retention of the labelled selenium.[10,11,13–16] Concern whether the tracer activity reflects the distribution of the native selenium has been raised. The specific incorporation of the labelled element into proteins has been found to be more effective the lower the Se status of the animal and the smaller the amount of tracer administered. In ecotoxicology, reliance on radiotracers is incompatible with the analysis of field-collected samples which is critical to the understanding of Se biogeochemistry and ecotoxicology.

A number of sensitive techniques such as instrumental neutron activation analysis (INAA),[15,17–19] proton induced X-ray emission (PIXE),[20] high resolution (HR) ICP MS,[21] or electrothermal vaporisation (ETV) AAS[13] have been proposed to determine the trace elements in the gel strips obtained after electrophoresis.[22] Microbeam X-ray spectrometry, energy-dispersive X-ray fluorescence spectrometry and INAA have been evaluated for the detection of Se in glutathione peroxidase previously separated by PAGE.[23] The possible role of ion-beam spectrometric analysis in protein analysis by PAGE or thin layer chromatography (TLC) has been discussed with emphasis on PIXE.[24,25]

A recent analytical technique is based on the sampling of a PAGE plate by laser ablation which is rapid and offers high (spatial) resolution.[8] Comparison of the Co distribution maps thereby obtained with protein distribution maps obtained by staining with Coomassie Brilliant Blue allow the identification of the main Co-binding proteins in serum.[8] The technique has been reported for speciation analysis of selenoproteins in Se-contaminated wildlife[9] and Pb complexes with humic and fulvic acids.[26] Applications of LA ICP MS in phosphoroproteomics are imminent.[27]

3 Capillary Zone Electrophoresis

Capillary zone electrophoresis (CZE) offers highly efficient separation, rapid analysis and minute sample size requirements. It is an established separation technique for metalloproteins and metal-binding peptides owing to the high separation efficiency, the small sample requirement (several nanoliters), and the absence of packing susceptible to interaction with metals and affecting the complexation equilibria.[28] These features have resulted in a number of applications of CZE for the separation of metallobiomolecules.[28] All this work has been carried out with UV detection which does not offer elemental specificity. A comprehensive review of reported methods concerning speciation of oxidation states, metal complexes with inorganic and organic ligands, and macromolecules such as humic substances and proteins is available.[29]

Principle of Separation by Capillary Zone Electrophoresis

In CZE the separation is achieved in an electrolyte-filled capillary. When an electric field (E) is applied across the capillary, ionic species migrate towards the oppositely charged electrode with a linear velocity $v_{\text{eff}} = \mu_{\text{eff}}\, E$, where μ_{eff} is the electrophoretic mobility of the species in the electrolyte. The use of ICP MS requires that the analyte be physically transported from the capillary to the ionisation source. This is possible owing to the electroosmotic flow.

In aqueous conditions, a fused silica capillary surface is covered by silanol groups, the ionisation of which depends on their pKa and on the pH of the buffer. At pH > 3, the negatively charged surface silanol groups are counterbalanced by positive ions from the buffer with a resulting formation of a double layer characterised by a zeta potential. The positive ions in the diffusion region migrate towards the cathode. Because of their solvation, the ion movement drags the bulk solution in the capillary towards the cathode resulting in an electroosmotic flow.

Figure 5.2 shows schematically the different movements involved in capillary zone electrophoresis. In the CZE-ICP MS coupling the species are driven to the nebuliser by electroosmotic flow. When the latter is larger than the electrophoretic migration, even negatively charged species can reach the detector. The separation of neutral species is possible by the creation of micelles, resulting in micellar electrokinetic chromatography (MEKC).

Interfacing Capillary Zone Electrophoresis to ICP MS

Attempts to use element selective detection for CZE turned out to be successful with a description of a CZE-ICP MS interface by Olesik *et al.*[30] Since then a number of interface designs have been reported. They are based on a (micro)-

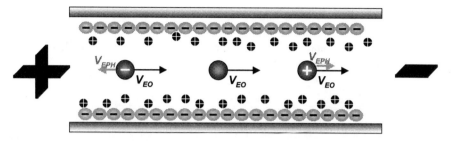

Figure 5.2 *Principle of capillary zone electrophporesis: V_{EO} – electroosmotic flow; V_{EPH} – electrophoretic velocity within the migration buffer*

concentric,[31–37] ultrasonic,[38] direct injection nebuliser,[39,40] or a high-efficiency nebuliser,[41] and use a coaxial electrolyte sheath flow (make up) to establish the electrical connection to the analyte. The principle and designs have been reviewed.[4,42,43] Sources of error in CZE-ICP MS coupling have been discussed.[44] Except for the work of Michalke[45] whose coupling was based on the collection of UV-monitored fractions eluting from the capillary and their analysis by ETV-ICP MS, the CZE-ICP MS interface has allowed the continuous detection of analytes. The problems relevant to interfacing CZE to ICP MS include:

(1) the apparently inadequate concentration detection limits to match the naturally occurring metallocompounds in a sample. Indeed, the minute quantity of sample injected, low transport efficiency and the sheath-flow dilution negatively affect the sensitivity. The dilution by the sheath-flow can be reduced by using a micronebuliser working at $1-10~\mu L\,min^{-1}$,[46–48] while the low transport efficiency can be remedied by the use of a direct injection,[39,40,49] or a total consumption micronebuliser.[46–48] Another way to improve the sensitivity is to use a sector field instrument but the gain over the new generation of ICP quadrupole MS instruments (*e.g.* Agilent 7500 series) seems to be minimal.

(2) the occurrence of a laminar flow through the capillary as a result of the nebuliser suction, and, consequently, a loss of the separation efficiency. The suction effect can be compensated by applying a negative pressure to the inlet vial or back pressure on the level of the make up flow. This, however, inadvertently results in sensitive fluid mechanical equilibria in the system and leads to the instability of migration time, separation efficiency and sensitivity. Interfaces using suction-free nebulisers such as direct injection (DIN),[39,40] or ultrasonic nebulisers,[38] have been reported but the analytical performance suffered from a $100-1000$-fold dilution of the electroosmotic flow (*ca.* $1~\mu L\,min^{-1}$) and from a higher background noise.[38] A self-aspirating sheath-flow interface technique has been proposed to reduce the suction effect of the nebuliser.[46,47,50,51] A direct injection high efficiency nebuliser operating at low uptake rates ($10~\mu L\,min^{-1}$) has recently been developed.[49]

In our opinion the two most attractive interface designs are those based on a modified MCN-100 nebuliser with a drain-free low-volume spray chamber[47] and on a direct injection nebuliser.[49] The former (the commercially available CEI-100 interface) together with a typical electropherogram is shown in Figure 5.3. A test

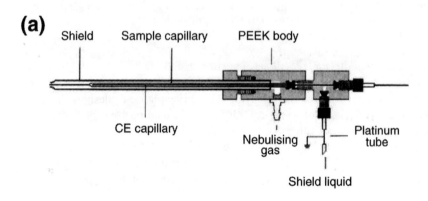

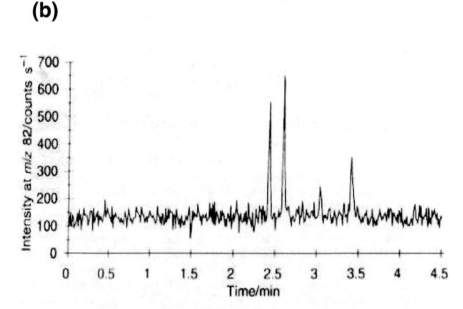

Figure 5.3 *Interfacing of capillary zone electrophoresis with ICP MS via direct injection nebuliser: a) scheme of the interface; (b) electropherogram obtained for a mixture of selenium species (0.5 µg Se L⁻¹ of each compound); elution order: selenate, selenite, selenocysteine and selenomethionine*
(from Ref. 49)

for a CZE-ICP MS proving the absence of the laminar flow is the comparison of the separation efficiency when a UV (on-capillary) and an ICP MS detector is used. A scheme of the interface based on the direct injection nebuliser together with a typical electropherogram is shown in Figure 5.4.

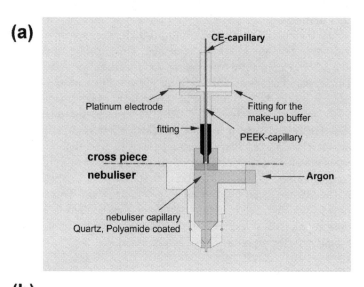

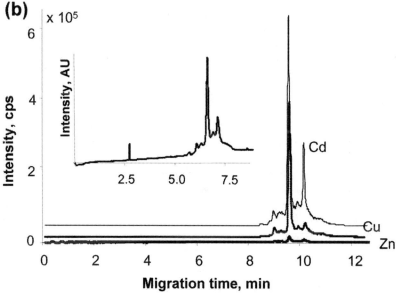

Figure 5.4 *Interfacing of capillary zone electrophoresis with ICP MS via a CEI-100 interface (CETAC): a) scheme of the interface (courtesy of Dr. Dirk Schaum-loeffel); b) electropherogram obtained from a mixture of metal complexes with MT-isoforms (CZE-UV electropherogram is shown in the inset)*

4 Areas of Application

PAGE has widely been used for the separation of selenoproteins (*cf.* Chapter 15) which remain stable during denaturation. Speciation studies of metal-protein complexes in which the metal-protein binding would have remained intact have been scarce.[5,6,8,17,19–21,52–54]

The application potential of CZE-ICP MS is large as demonstrated by the reviews on CZE of metal species,[29] CZE of Pt-group elements,[55] and CZE of metalloproteins and metal-binding peptides.[28] In practice, however, applications of CZE-ICP MS to speciation analysis have been rare. Only in few cases was this coupling capable of producing valid information for speciation of an element in a real sample. They concerned speciation of Se in human milk and blood serum[56] and that of iodine in milk.[57] Some authors have used the metallothionein standards[31,33,37,38,44] but both the number of these papers and the quality of the electropherograms reported were incredibly low in comparison with the data available in the literature for capillary electrophoresis of metallothionein.[58] The notable exceptions are studies on speciation of metal complexes with metallothioneins in rat liver and kidney[59,60] and human brain tissues.[61]

Quadrupole ICP MS has been used in all these studies with the exception of one recent study in which a sector field instrument was applied.[46] The use of collision cell quadrupole MS detection for the determination of monophosphate nucleotides by capillary electrophoresis has been reported.[62]

References

1. R. Lobinski and M. Potin Gautier, *Analysis*, 1998, **26**, M21.
2. A. Sanz-Medel, *Spectrochim. Acta*, 1998, **53B**, 197.
3. J. Szpunar, *Analyst*, 2000, **125**, 963.
4. R. M. Barnes, *Fresenius' J. Anal. Chem.*, 1998, **361**, 246.
5. S. Lustig, J. De Kimpe, R. Cornelis and P. Schramel, *Fresenius' J. Anal. Chem.*, 1999, **363**, 484.
6. S. Lustig, J. De Kimpe, R. Cornelis, P. Schramel and B. Michalke, *Electrophoresis*, 1999, **20**, 1627.
7. J.L. Nielsen, O.M. Poulsen and A. Abildtrup, *Electrophoresis*, 1994, **15**, 666.
8. J.L. Nielsen, A. Abildtrup, J. Christensen, P. Watson, A. Cox and C.W. McLeod, *Spectrochim. Acta*, 1998, **53B**, 339.
9. T.W.M. Fan, E. Pruszkowski and S. Shuttleworth, *J. Anal. At. Spectrom.*, 2002, **17**, 1621.
10. D. Behne, C. Weiss Nowak, M. Kalcklosch, C. Westphal, H. Gessner and A. Kyriakopoulos, *Analyst*, 1995, **120**, 823.
11. D. Behne, C. Hammel, H. Pfeifer, D. Rothlein, H. Gessner and A. Kyriakopoulos, *Analyst*, 1998, **123**, 871.
12. D. Behne, H. Hilmert, S. Scheid, H. Gessner and W. Elger, *Biochim. Biophys. Acta*, 1988, **966**, 12.
13. A. Kyriakopoulos, M. Kalcklosch, C. Weiss-Nowak and D. Behne, *Electrophoresis*, 1993, **14**, 108.

14. D. Behne, A. Kyriakopoulos, M. Kalckosch, C. Weiss-Nowak, H. Pfeifer, H. Gessner and C. Hammel, *Biomed. Environ. Sci.*, 1997, **10**, 340.
15. D. Behne, S. Scheid, H. Hilmert, H. Gessner, D. Gawlik and A. Kyriakopoulos, *Biol. Trace Element Res.*, 1990, **26–27**, 439.
16. C.C. Chery, E. Dumont, R. Cornelis and L. Moens, *Fresenius J. Anal. Chem.*, 2001, **371**, 775.
17. S.F. Stone, D. Hancock and R. Zeisler, *J. Radioanal. Nucl. Chem.*, 1987, **112**, 95.
18. C.K. Jayawickreme and A. Chatt, *J. Radioanal. Nucl. Chem.*, 1988, **124**, 257.
19. S.F. Stone, R. Zeisler and G.E. Gordon, *Biol. Trace Element Res.*, 1990, **26–27**, 85.
20. Z. Szokefalvi Nagy, I. Demeter, C. Bagyinka and K.L. Kovacs, *Nucl. Instrum. Meth. Phys. Res.*, 1987, **B22**, 156.
21. S. Lustig, D. Lampaert, K. De Cremer, J. De Kimpe, R. Cornelis and P. Schramel, *J. Anal. At. Spectrom.*, 1999, **14**, 1357.
22. C.C. Chery, H. Chassaigne, L. Verbeeck, R. Cornelis, F. Vanhaecke and L. Moens, *J. Anal. At. Spectrom.*, 2002, **17**, 576.
23. S.F. Stone, G. Bernasconi, N. Haselberger, M. Makarewicz, R. Ogris, R. Wobrauschek and R. Zeisler, *Biol. Trace Elem. Res.*, 1994, **43–45**, 299.
24. Z. Szokefalvi Nagy, *Nucl. Instrum. Meth. Phys Res.*, 1996, **B238**, 110234.
25. Z. Szokefalvi Nagy, C. Bagyinka, I. Demeter, K. Hollos Nagy and K.L. Kovacs, *Fresenius' J. Anal. Chem.*, 1999, **363**, 469.
26. R.D. Evans and J.Y. Villeneuve, *J. Anal. At. Spectrom.*, 2000, **15**, 157.
27. M. Wind, I. Feldmann, N. Jakubowski, W.D. Lehmann, *Electrophoresis*, 2003, **24**, 1276.
28. M.P. Richards and J.H. Beattie, *J. Cap. Elec.*, 1994, **3**, 196.
29. E. Dabek Zlotorzynska, E.P.C. Lai and A.R. Timerbaev, *Anal. Chim. Acta*, 1998, **359**, 1.
30. J.W. Olesik, J.A. Kinzer and S.V. Olesik, *Anal. Chem.*, 1995, **67**, 1.
31. Q. Lu, S.M. Bird and R.M. Barnes, *Anal. Chem.*, 1995, **67**, 2949.
32. M. Van Holderbeke, Y.N. Zhao, F. Vanhaecke, L. Moens, R. Dams and P. Sandra, *J. Anal. At. Spectrom.*, 1999, **14**, 229.
33. S.A. Baker and N.J. Miller Ihli, *Appl. Spectrosc.*, 1999, **53**, 471.
34. V. Majidi and N.J. Miller Ihli, *Analyst*, 1998, **123**, 803.
35. B. Michalke and P. Schramel, *Analysis*, 1998, **26**, M51.
36. B. Michalke and P. Schramel, *Fresenius' J. Anal. Chem.*, 1997, **357**, 594.
37. K.A. Taylor, B.L. Sharp, D.J. Lewis and H.M. Crews, *J. Anal. At. Spectrom.*, 1998, **13**, 1095.
38. Q.H. Lu and R.M. Barnes, *Microchem. J.*, 1996, **54**, 129.
39. Y. Liu, V. Lopez Avila, J.J. Zhu, D.R. Wiederin and W.F. Beckert, *Anal. Chem.*, 1995, **67**, 2020.
40. A. Tangen, W. Lund, B. Josefsson and H. Borg, *J. Chromatogr A*, 1998, **826**, 87.
41. K.L. Sutton, C. B'Hymer and J.A. Caruso, *J. Anal. At. Spectrom.*, 1998, **13**, 885.
42. K.L. Sutton and J.A. Caruso, *LC GC*, 1999, **17**, 36.

43. J.W. Olesik, J.A. Kinzer, E.J. Grunwald, K.K. Thaxton and S.V. Olesik, *Spectrochim. Acta*, 1998, **53B**, 239.
44. V. Majidi and N.J. Miller Ihli, *Analyst*, 1998, **123**, 809.
45. B. Michalke and P. Schramel, *J. Chromatogr A*, 1996, **750**, 51.
46. A. Prange and D. Schaumlöffel, *J. Anal. At. Spectrom.*, 1999, **14**, 1329.
47. D. Schaumlöffel and A. Prange, *Fresenius' J. Anal. Chem.*, 2000, **364**, 452.
48. K. Polec, J. Szpunar, O. Palacios, P. Gonzalez-Duarte, S. Atrian and R. Lobinski, *J. Anal. At. Spectrom.*, 2001, **16**, 567.
49. L. Bendahl, B. Gammelgaard, O. Jons, O. Farver and S.H. Hansen, *J. Anal. At. Spectrom.*, 2001, **16**, 38.
50. C. B'Hymer, J.A. Day and J.A. Caruso, *Appl. Spectrosc.*, 2000, **54**, 1040.
51. J.A. Day, K.L. Sutton, R.S. Soman and J.A. Caruso, *Analyst*, 2000, **125**, 819.
52. Z.B. Szokefalvi-Nagy, C. Bagyinka, I. Demeter, K.L. Kovacs, L.H. Quynh, *Biol. Trace Element. Res.*, 1990, **26–27**, 93.
53. Y. Makino and E. Kawanishi, *J. Chromatogr, Biomed Appl.*, 1991, **105**, 248.
54. L. Dunemann and H. Reinecke, *Fresenius' J. Anal. Chem.*, 1989, **334**, 743.
55. A.R. Timerbaev, A. Küng and B.K. Keppler, *J. Chromatogr. A*, 2002, **945**, 25.
56. B. Michalke and P. Schramel, *J. Chromatog. A*, 1998, **807**, 71.
57. B. Michalke, *J. Anal. At. Spectrom.*, 1999, **14**, 1297.
58. R. Lobinski, H. Chassaigne and J. Szpunar, *Talanta*, 1998, **46**, 271.
59. K. Polec, M. Peréz-Calvo, O. Garcia-Arribas, J. Szpunar, B. Ribas-Ozonas and R. Lobinski, *J. Inorg. Biochem.*, 2001, **88**, 197.
60. K. Polec-Pawlak, D. Schaumloeffel, J. Szpunar, A. Prange and R. Lobinski, *J. Anal. At. Spectrom.*, 2002, **17**, 908.
61. D. Schaumloeffel, A. Prange, G. Marx, K.G. Heumann and P. Braetter, *Anal. Bioanal. Chem.*, 2002, **372**, 155.
62. C.F. Yeh and S.J. Jiang, *Analyst*, 2002, **127**, 1324.

Electrospray Mass Spectrometry in Elemental Speciation Analysis

1 Introduction

Atomic spectrometric detection in chromatography is element specific. The species selectivity is a function of the arrival time of the analyte molecule at the ionisation source. Consequently, the identification of the eluted compound requires a prior knowledge of the possible species present and the availability of retention or migration time standards. The shift of interest in elemental speciation from well-defined organometallic anthropogenic contaminants to endogenous metal species of natural origin has created a need for the identification of analytes, many of which have not yet been reported, and for most of which retention time standards are not available.

The access to structural information for the identification of novel compounds is a great challenge in speciation analysis, especially as the improving sensitivity of ICP MS instruments will inevitably increase the number of metal and metalloid species detected. Potential opportunities for the accurate determination of the molecular weight and structural characterisation of molecules at trace levels in fairly complex matrices are offered by electrospray mass spectrometry (ES MS) and matrix assisted laser desorption ionisation (MALDI)-TOF MS. ES MS has become a detection technique complementary to ICP MS in HPLC and CZE which allows an accurate determination of the molar mass of chemical species and *via* collision induced dissociation of the protonated molecular ion, acquisition of structural information.[1,2] The use of MALDI-TOF MS for this purpose is in its infancy.[3]

At the present stage of development the sensitivity of ES MS still remains about two orders of magnitude lower than that of ICP MS and the vulnerability of the technique to the matrix (especially if rich in salts) requires a number of precautions to be taken for successful analysis. They include purification of crude extracts by multidimensional chromatography and desalting of the fractions of interest. These protocols are similar to those used prior to NMR and Edman degradation studies but for ES MS and MALDI MS they can be considerably scaled down requiring 100–1000 times less analyte with a much less critical degree of purification.

Useful mass spectra can be obtained only from relatively stable metallospecies with a covalent metal-carbon bond or complexes such as cobalamins, porphyrins or metalloproteins. The metal-ligand bond in labile complexes is likely to be destroyed during the ionisation process. In such cases the ligand alone should be identified and characterised by ES MS/MS whereas the existence of the metal-ligand link should be demonstrated by a complementary analytical technique.

The progress in electrospray MS/MS in elemental speciation analysis is accelerating with a wider availability of triple quadrupole and Q-TOF instruments, the latter allowing accurate molecular mass measurements, increased sensitivity and improved tolerance to the matrix. The various facets of electrospray MS in speciation analysis have been reviewed.[1,2]

2 Principles of Electrospray Mass Spectrometry

Electrospray Ionisation

Particular attention in the field of analysis for metal species has been paid to electrospray (ES) ionisation and its pneumatically assisted modification referred to as Ion-spray®. The technique has been shown to be capable of producing gas-phase ions of highly labile and non-volatile compounds; in particular metal complexes of Cd, Zn and Ag have been shown to be transferred into the gaseous phase and detected by the mass spectrometer.[4–7]

The principle of electrospray ionisation is schematically shown in Figure 6.1. It consists of pumping a solution containing the analytes at a rate of $1-10~\mu\text{L min}^{-1}$

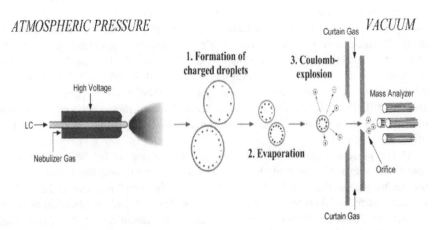

Figure 6.1 *Principle of electrospray ionisation* (adapted from Applied Biosystems)

through a narrow bore stainless steel or fused silica capillary that is biased to a high potential relative to a ground planar electrode, referred to as the counter electrode. Ions in solutions are able to evaporate from the small, highly charged droplets generated upon nebulisation and subsequent evaporation of the neutral solvent molecules. The ions formed are transported from the atmospheric pressure area to a high vacuum where they are subject to separation as a function of their mass-to-charge ratio using a mass analyser. A particular feature of electrospray ionisation is the formation of multiple charged ions from peptides and proteins.

Mass Spectrometric Analysers

The principal types of mass analysers used in electrospray MS for speciation analysis include quadrupole, ion-trap, time-of-flight and Fourier Transform ion cyclotron resonance. They are shown schematically in Figure 6.2.

A quadrupole mass analyser (Figure 6.2a) has been the most frequently used. It offers a unit resolution at which the isotopic cluster of doubly charged ions is sufficiently resolved that monoisotopic masses are measured. The resolution is insufficient for ions that are triply (or more) charged so that the average mass of a polypeptide is measured. The accuracy of the m/z measurement is within 0.1 Da. The unit resolution of a quadrupole filter is insufficient for the unambiguous confirmation of the compound identity on the basis of the molecular mass. A quadrupole is a sequential scanning analyser so the majority of ions are purposefully directed out of the instrument and cannot be detected. In the case of a 1.8 s long scan from 200 to 2000, only 0.001 s is spent at each m/z. Because all ions are produced continuously, this means that only 0.1% of the ions produced are detected which results in a considerable decrease in sensitivity in the scanning mode.

An ion trap mass analyser (Figure 6.2b) offers improved sensitivity owing to the compact size of the analyser (improved transmission) and utilisation of a larger proportion of the ions that are formed by the ion source. In the first step ions formed in the source are accumulated. Subsequently the ion of interest is trapped whereas the others are ejected. All the ions injected into the trap can be ultimately sent to the detector. Mass resolution is similar to that in quadrupole mass filters. A consequence of the higher sensitivity is, however, that the resolution does not need to be sacrificed and the full mass resolution capabilities are used. A particular feature of an ion trap mass spectrometer is the possibility of refragmenting the fragments of the molecular ion, resulting in the acquisition of MS^n spectra that can be used for structure elucidation of unknown compounds (*cf.* Chapter 14).

Better precision and higher mass measurement accuracy can be obtained by using a TOF mass analyser (Figure 6.2c). It uses the differences in transit time through a drift region to separate ions of different masses. An electric field accelerates all ions into a field-free drift region with a kinetic energy of qV, where q is the ion charge and V is the applied voltage. Since the kinetic energy of the ion is expressed as $0.5mv^2$, lighter ions have a higher velocity than heavier ones

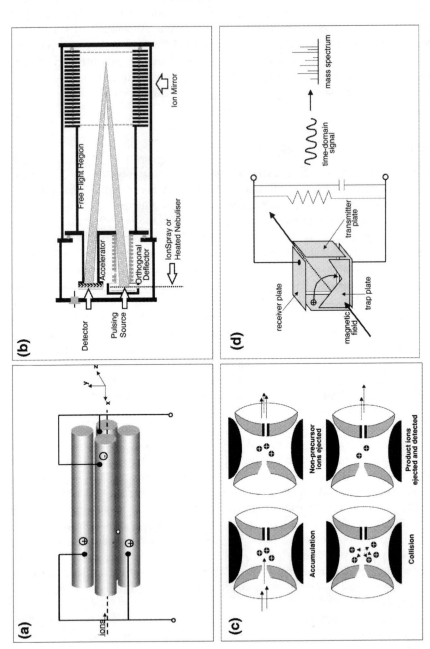

Figure 6.2 *Types of mass spectrometric analysers: (a) quadrupole; (b) time-of-flight; (c) ion trap; (d) Fourier Transform ion cyclotron resonance*

and reach the detector installed at the end of the drift region sooner. The combination of time delayed extraction of ions with the use of an ion reflectron allows resolution above 10 000 be obtained with a mass accuracy below 10 ppm.

A more accurate confirmation can be obtained by improving the resolution, which is possible by a Fourier Transform ion cyclotron resonance mass spectrometer (Figure 6.2d). In this technique ions entering a chamber are trapped by a powerful magnetic field in circular orbits with a frequency independent of their velocity. The cyclotron motion of ions is excited by a RF field to generate a time dependent current. The current is converted by FT into orbital frequencies of the ions which correspond to their mass-to-charge ratios. In this way resolution exceeding 200 000 can be obtained and mass accuracy down to 0.2 ppm.

Tandem Mass Spectrometry

In tandem mass spectrometry an ion selected by the first mass analyser undergoes fragmentation by collision with inert gas molecules in a collision cell and the fragments are further analysed. Tandem mass spectrometers are divided into:

(1) tandem in space mass spectrometers in which the selection of the ion to be fragmented and analyses of the fragments are carried out in two mass analysers connected by a collision cell (QqQ and QqTOF).
(2) tandem in time mass spectrometers in which the selection of the ion, its fragmentation and analyses of fragments are carried out in the same place but are separated in time (ion trap and ICR instruments).

The use of tandem mass spectrometry for speciation analysis is rapidly gaining momentum. In most of the cases the product ion scan mode is selected to obtain structural information from ions produced in the ion source. It consists of the selection of the protonated molecular ion of the species of interest, fragmention in a collision cell and the determination of the molecular masses of the fragments. When an ion trap spectrometer is used the fragment scan be refragmented allowing structural information on the product ions to be obtained.

On-Line Coupling of Electrospray MS to Separation Techniques

The major advantage of ES MS over the other soft MS techniques is the possibility of its application as a chromatographic or an electrophoretic detector. The prerequisite is to match the optimum ionisation conditions with the optimum separation conditions in terms of the chemical composition and flow rate of the mobile phase. Whereas different developments with regard to the source (nanospray, micro ion spray or turbo ion spray) have allowed different flow rates (from nL min^{-1} to mL min^{-1}) to be handled, the performance of ES MS remains negatively affected by salt concentrations exceeding 10 mM and by the presence of some compounds, such as ion pairing reagents, even if present at the mM level.

Time-of-flight MS is an ideal detector for LC and CZE because full scans of a relatively large m/z region can be obtained at a high frequency allowing acquistion of mass spectra at any moment in the chromatographic elution with high sensitivity. Quadrupole mass analysis may be the technique of choice when the molecular mass of the species to be determined is known. When a triple quadrupole analyser is available multiple reaction mode (monitoring of the formation of a particular fragment from a predefined molecular ion) offers a particularly high signal-to-background ratio and thus high measurement sensitivity.

3 Speciation-Relevant Information from Electrospray MS

Depending on how the generated ion is further handled, different types of information regarding the identity of the molecule of interest can be obtained. In the simplest case (molecular ion or single MS mode), one or more protons can be attached to the analyte molecule leading to the formation of a singly (generally for species up to 1000 u) or multiplycharged ion. This allows an accurate measurement of the molecular mass of the analyte species of interest. The protonated molecule ions $(M + nH^+)^{n+}$ can be broken down by increasing the ionisation energy (orifice voltage) at the source in the process referred to as source collision induced dissociation (SCID).[8,9] At a sufficiently high ionisation energy elemental ions of metals or metalloids present in the sample can be obtained.[8,9] Finally, the molecular ion, isolated at the level of the first mass filter can be fragmented by collisions with molecules of a neutral gas in a process referred to as collision induced dissociation (CID). The resulting fragments identified as to molecular masses by a second mass analyser (tandem MS or MS/MS mode) allow information on the structure of the molecule to be obtained. Typical mass spectra obtained in the different modes are shown in Figure 6.3.

Molecular Mass Determination Based on a Singly Charged Molecular Ion

Figure 6.3a shows an ESI mass spectrum obtained for a common organoselenium compound in biological samples – selenomethionine. It shows two m/z envelopes within which the isotopic pattern of selenium can be identified. The spectrum allows the identification of a selenium compound with the molecular mass of 198.0 (for the most abundant ^{80}Se isotope). Despite the relatively low energy (orifice potential of 20 V) a fragment (M-17) corresponding to the loss of an OH group from the carboxylic group of the selenoamino acid can be observed.

The effect of the accuracy of the mass analyser (quadrupole, TOF or FT ICR) used on the reliability of the determination of species containing a monoisotopic heteroelement, such as arsenic, is discussed in more detail in Chapter 14.

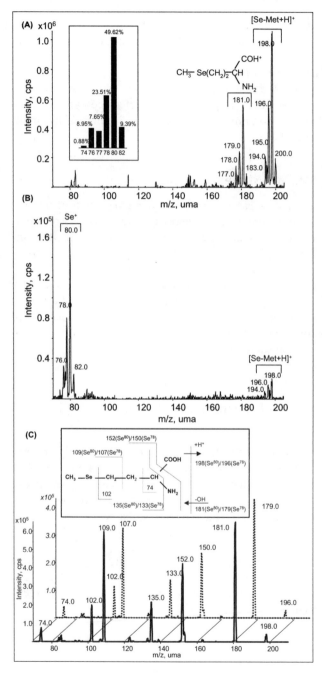

Figure 6.3 *Electrospray mass spectra of selenomethionine: (a) molecular; (b) elemental; (c) tandem mode*
(Reprinted from *Trends Anal. Chem.*, 2000, **19**, 302, copyright 2000, with permission from Elsevier)

Elemental (SCID) Mode

The application of a ten-fold higher orifice potential (200 V) results in the destruction of the molecular ion and leads to the formation of the elemental Se cation (Figure 6.3b). This can be of benefit in the case of real life samples when a number of concomitant ions can interfere with the identification of the Se isotopic pattern at m/z 198.0 at low ionisation potentials. Note also that the elemental ESI MS mode is free from Ar polyatomic interferences present in ICP MS. The concentration detection limits, however, are 2–3 orders of magnitude higher than in the case of ICP MS.

Tandem MS/MS Mode

Structural information can be obtained by fragmentation of the molecular ion as demonstrated in Figure 6.3c. In particular, for species containing an element having more than one stable isotope, such as selenium, valuable information can be obtained by fragmenting the two protonated molecule ions containing the adjacent most abundant isotopes (^{78}Se and ^{80}Se).[10] Fragments that contain selenium will still be separated by a distance of two units whereas fragments that do not will remain at the same m/z value thus facilitating the interpretation of the mass spectra. The fragmentation pattern is usually rather rich; practically, ions of all the possible fragments are represented in the mass spectrum. Note, however, that the atomic ion cannot be seen, the CID conditions being inappropriate for its generation.

Molecular Mass Determination Based on a Multiply Charged Molecular Ion

An attractive feature of ESI MS is the formation of multiply protonated molecular ions from polypeptide molecules and their complexes with metals. This allows a very precise determination of the molecular mass of the species of interest. Assuming that adjacent peaks in the ion envelope differ by only one charge and that the charge is due to protonation, the relation between a multiply charged peak at $m/z = p_1$ and the relative molecular mass M_r is:

$$p_1 z_1 = M_r + m_H z_i \tag{1}$$

where m_H is the mass of the proton and z_i is the number of charges. The adjacent peak at the high m/z side of p_1 is described by:

$$p_2(z_i - 1) = M_r + m_H(z_i - 1) \tag{2}$$

Hence $z_i = (p_2 - m_H)/(p_2 - p_1)$ and the molecular weight can be calculated from Equation 1. The higher the number of m/z signals that can be seen for a given compound in the mass spectrum, the more precise the result of the molecular mass determination. Computer algorithms are available to increase the precision of the molecular mass determination when more than two peaks corresponding to different charge states are present in the mass spectrum.

Figure 6.4 shows an ESI MS spectrum for a Cd complex with metallothionein isolated from horse kidney. Four envelopes of multicharged peaks can be identified. The attribution of two adjacent ionisation states to the same molecule allows the accurate determination (+0.1%) of the molecular mass of each compound present. It can hence be demonstrated that except for the Cd-MT complex the preparation contained three other proteins with molecular masses equal to 8180.5 ± 0.5, 8295.0 ± 0.0 and 12005.4 ± 0.8 Da.

Sequencing of Oligopeptides by ESI Tandem MS

Peptides of limited molecular weight (up to 2.5 kDa) can be subject to on-line sequence analysis by MS/MS using CID.[11] Peptides fragment primarily at the amine bonds to produce a ladder of sequence ions. The charge can be retained on the amino terminus (type b ion) or on the carboxy terminus (type y ion). Thus a complete series made up of ions from both types allows the determination of the amino acid sequence by subtraction of the masses of adjacent sequence ions. Figure 6.5 demonstrates the method as applied to a selenopeptide ($M_r = 595$ Da). Most of the expected b- and y-type fragments can be identified.

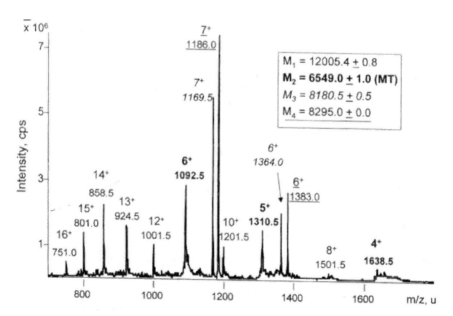

Figure 6.4 *Determination of the molecular mass of a protein using the multiply charged ion envelope. Four proteins are present: a) $M_r = 12\,005.4 + 0.8$ (normal font); b) $M_r = 6549.0 + 1.0$ (horse kidney $MT - Cd_4$ complex) (bold font); c) $M_r = 8180.5 + 0.5$ (italics); d) $M_r = 8295.0 + 0.0$ (underlined)*
(Reprinted from *Trends Anal. Chem.*, 2000, **19**, 302, copyright 2000, with permission from Elsevier))

(a)

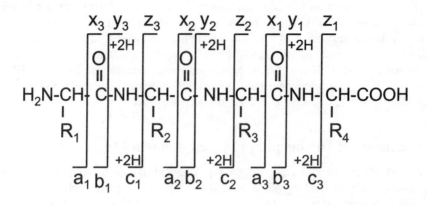

(b)

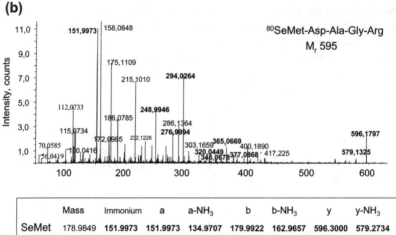

	Mass	Immonium	a	a-NH$_3$	b	b-NH$_3$	y	y-NH$_3$
SeMet	178.9849	151.9973	151.9973	134.9707	179.9922	162.9657	596.3000	579.2734
Asn	114.0429	87.0553	266.0402	249.0137	294.0351	277.0086	417.3150	400.2885
Ala	71.0371	44.0495	337.0773	320.0508	365.0723	348.0457	303.2721	286.2456
Gly	57.0215	30.0338	394.0988	377.0723	422.0937	405.0672	232.2350	215.2085
Arg	156.1011	129.1135	550.1999	533.1734	578.1948	561.1683	175.2135	158.1870

Figure 6.5 *Sequencing of polypeptides by collision induced dissociation MS: a) principle (adapted from Ref. 11); b) application to a selenopeptide. Numbers marked in bold refer to m/z of fragments which were found in the spectrum*

4 Areas of Application

Identification of Organometallic (Metalloid) Species by ES MS

Small molecules containing a carbon-metal (metalloid) bond usually produce readily singly protonated ions in the electrospray source which theoretically should allow the identification of the metallocompound on the basis of the molecular mass. However, in the direct infusion mode, the attribution of a signal at a given m/z ratio to an elemental species is a daunting and practically impossible task for monoisotopic elements such as arsenic. Sharp *et al.*[12] stated that ionspray MS is sensitive to virtually all the species in a sample and the resulting spectra for real samples are too complex, too likely to contain multiply charged species and too prone to artefact formation in the gas phase to be suitable for speciation without prior chromatography or electrophoresis. However, if an element presents a characteristic isotopic pattern such as Cd,[13] Se[10,14–17] or Sn,[18,19] its recognition in the mass spectrum of a sample solution is easier provided that the signal is not suppressed by the matrix.

The MS/MS mode has allowed the identification of organoarsenic compounds in algal and clam tissue extracts,[20,21] organoselenium in yeast[10,22] and garlic[23] and of the Al-citrate complex in human serum.[24]

HPLC with on-line ES MS detection has been reported for organoarsenic species[25–27] and organoselenium.[15,16,28] Since a quadrupole mass spectrometer has been used in these studies the prerequisite of success was the knowledge of at least the molecular mass of the analysed species. Also the MS/MS mode has allowed the identification of As compounds in HPLC effluents.[25,29] CZE-ES MS has been shown to identify arsenocholine in urine but the identification (by spiking and migration time matching) does not seem to be convincing in view of the ease of the formation of artefacts in the protonated molecule ion mode (no CID MS spectra were acquired).[22]

Characterisation of Metal Complexes with Peptides and Proteins

Owing to the formation of ions stable in the gas phase even by metallated metallothioneins and to the very precise (0.1%) determination of the molecular mass of the species of interest, information can be obtained about the number and identity of metals bound to the protein from characteristic spectra for apoproteins and metal-saturated proteins at acidic and alkaline pH. A single measurement provides information about molar distributions and estimates of the relative abundances of various complexes in the sample.

Several studies have examined the characterisation of metallothionein (MT) complexes with Cd and Zn by microbore HPLC[5,6,30–32] and CZE[7,33] with ES MS detection. RP HPLC of metallated isoforms is faster but prone to artefacts because of a number of mixed metal complexes present.[32] A remedy for this is post-column acidification which cleaves the metals bound to MT and allows both information on the identity of the bound metal[6] and on that of the MT-ligand present to be obtained. The resolution of CZE allows the separation of metal-free MTs.[7,33]

The use of CID MS for the identification of phytochelatin ligands in phytochelatin (PC)-metal complexes detected by HPLC-ICP MS has been demonstrated.[34,35]

There is continual interest in the CZE-ES MS/MS coupling for structural characterisation of organic molecules but for the moment the poor detection limits (also due to the highly conductive mobile phases) have practically prohibited its application to bioinorganic trace element speciation analysis. Nevertheless a number of reports related to the analysis of metallothionein-metal complexes[7,33,36] and for speciation analysis in standard solutions do exist.[37,38] CZE-MS/MS has been proposed for the characterisation (including sequencing of phytochelatins).[39]

References

1. D.A. Barnett, R. Handy and G. Horlick, in J.A. Caruso, K.L. Sutton and K.L. Ackley, (Eds.), *Elemental Speciation. New Approaches for Trace Element Analysis,* Elsevier, Amsterdam, 2000.
2. H. Chassaigne, V. Vacchina and R. Lobinski, *Trends Anal. Chem.*, 2000, **19**, 300.
3. J. Ruiz Encinar, R. Ruzik, W. Buchmann, J. Tortajada, R. Lobinski and J. Szpunar, *Analyst*, 2003, **128**, 220.
4. X. Yu, M. Wojciechowski and C. Fenselau, *Anal. Chem.*, 1993, **65**, 1355.
5. H. Chassaigne and R. Lobinski, *Anal. Chem.*, 1998, **70**, 2536.
6. J.C.Y. Le Blanc, *J. Anal. At. Spectrom.*, 1997, **12**, 525.
7. X. Guo, H.M. Chan, R. Guevremont and K.W.M. Siu, *Rapid Commun. Mass Spectrom.*, 1999, **13**, 500.
8. J.J. Corr and J.F. Anacleto, *Anal. Chem.*, 1996, **68**, 2155.
9. J.J. Corr, *J. Anal. At. Spectrom.*, 1997, **12**, 537.
10. C. Casiot, V. Vacchina, H. Chassaigne, J. Szpunar, M. Potin-Gautier and R. Lobinski, *Anal. Commun.*, 1999, **36**, 77.
11. K. Biemann, *Annu. Rev. Biochem.*, 1992, **61**, 977 .
12. B.L. Sharp, A.B. Sulaiman, K.A. Taylor and B.N. Green, *J. Anal. At. Spectrom.*, 1997, **12**, 603.
13. K.A. High, B.A. Methven, J.W. McLaren, K.W.M. Siu, J. Wang, J.F. Klaverkamp and J.S. Blais, *Fresenius' J. Anal. Chem.*, 1995, **351**, 393.
14. H.M. Crews, P.A. Clarke, D.J. Lewis, L.M. Owen, P.R. Strutt and A. Izquierdo, *J. Anal. At. Spectrom.*, 1996, **11**, 1177.
15. M. Kotrebai, M. Biringer, J.F. Tyson, E. Block and P.C. Uden, *Anal. Commun.*, 1999, **36**, 249.
16. M. Kotrebai, J.F. Tyson, E. Block and P.C. Uden, *J. Chromatogr A*, 2000, **866**, 51.
17. T.M. Fan, A.N. Lane, D. Martens and R.M. Higashi, *Analyst*, 1998, **123**, 875.
18. T.L. Jones and L.D. Betowski, *Rapid Commun. Mass Spectrom.*, 1993, **7**, 1003.
19. K.W.M. Siu, G.J. Gardner and S.S. Berman, *Rapid Commun. Mass Spectrom.*, 1988, **2**, 201.
20. S. McSheehy, M. Marcinek, H. Chassaigne and J. Szpunar, *Anal. Chim. Acta*, 2000, **410**, 71.
21. S. McSheehy, J. Szpunar, R. Lobinski, V. Haldys, J. Tortajada and J. Edmonds, *Anal. Chem.*, 2002, **74**, 2370.

22. S. McSheehy, V. Haldys, J. Tortajada and J. Szpunar, *J. Anal. At. Spectrom.*, 2002, **17**, 507.
23. S. McSheehy, W. Yang, F. Pannier, J. Szpunar, R. Lobinski, J. Auger and M. Potin-Gautier, *Anal. Chim. Acta*, 2000, **421**, 157.
24. T. Bantan, R. Milacic, B. Mitrovic and B. Pihlar, *J. Anal. At. Spectrom.*, 1999, **14**, 1743.
25. J.J. Corr and E.H. Larsen, *J. Anal. At. Spectrom.*, 1996, **11**, 1215.
26. A.D. Madsen, W. Goessler, S.N. Pedersen and K.A. Francesconi, *J. Anal. At. Spectrom.*, 2000, **15**, 657.
27. S.N. Pedersen and K.A. Francesconi, *Rapid Commun. Mass Spectro.*, 2000, **14**, 641.
28. M. Kotrebai and P.C. Uden, *Spectrochim. Acta,* 1999, **54B**, 1573.
29. A. Le Bouil, A. Cailleux, A. Turcant and P. Allain, *J. Anal. Toxicol.*, 1999, **23**, 257.
30. H. Chassaigne and R. Lobinski, *Analyst*, 1998, **123**, 2125.
31. H. Chassaigne and R. Lobinski, *J. Chromatogr A*, 1998, **829**, 127.
32. H. Chassaigne and R. Lobinski, *Talanta*, 1999, **48**, 109.
33. C.B. Knudsen, I. Bjoernsdottir, O. Joens and S.H. Hansen, *Anal.Biochem.*, 1998, **265**, 167.
34. V. Vacchina, K. Polec and J. Szpunar, *J. Anal. At. Spectrom.*, 1999, **14**, 1557.
35. V. Vacchina, H. Chassaigne, M. Oven, M.H. Zenk and R. Lobinski, *Analyst*, 1999, **124**, 1425.
36. H. Chassaigne and R. Lobinski, *Fresenius' J. Anal. Chem.*, 1998, **361**, 267.
37. R.M. Barnes, *Fresenius' J. Anal. Chem.*, 1998, **361**, 246.
38. O. Schramel, B. Michalke and A. Kettrup, *J. Anal. At. Spectrom.*, 1999, **14**, 1339.
39. S. Mounicou, V. Vacchina, J. Szpunar, M. Potin-Gautier and R. Lobinski, *Analyst*, 2002, **126**, 624.

CHAPTER 7

Quality Control and Assurance in Speciation Analysis

1 Introduction

Speciation analysis, as any other field of analytical chemistry, requires quality control and assurance protocols to be implemented in order to produce valid data. The first step to achieve a valid speciation analysis is the definition of the target analyte. Many elemental species, especially redox states and coordination complexes, are inherently unstable and may change during storage or upon extraction from a matrix which has had a stabilising effect on the species.

Validation problems related to speciation of organometallic anthropogenic contaminants have been comprehensively discussed.[1] Several interlaboratory 'round-robin' exercises have resulted in an awareness of the problem and have led to solutions resulting in the availability of certified reference materials for organomercury, -tin, -lead and -arsenic species. In contrast the problem of validation of speciation analyses of endogenous metal species in biological materials is still open. The chromatographic purity of peaks and the identity of the species registered in chromatography often remain unknown. Also, it is often unknown whether the species arriving at the detector existed in the original sample or were simply created during the analytical procedure, *e.g.* by oxidation and/or ligand exchange.

Elemental speciation analysis by hyphenated techniques is a relatively young field and validation problems are still considered secondary. However, traceability is a prerequisite if speciation related measurements are to form the basis for decisions regarding, environmental policy or food quality regulations. This chapter briefly discusses issues related to the validation of speciation analyses, focusing on the analytical measurement.

2 Definition of the Target Moiety

The definition of a target species (moiety) is often problematic because of the complexity of real life systems. For example, the speciation of mercury in biological tissues is usually understood to be the differentiation between CH_3Hg^+ and Hg^{2+}. In reality, however, these species hardly ever exist as ions. Both methylmercury and Hg^{2+} can form covalent compounds with a counterion, coordination complexes with SH amino acids and bind strongly to cysteine residues of metallothioneins or larger proteins. On their turn, metallothioneins can polymerise to form dimers and higher oligomers each of which can bind (methyl)mercury. This is why the analytical target should be defined *à priori* in terms of our needs, usually resulting from eco- or clinical toxicology.

The target species should also be considered in terms of their thermodynamic stability and kinetic inertness. The timescale of the analytical method should match the lifetime of the target. Although metal ions can interact with many organic ligands, labile coordination compounds cannot be separated or analysed directly. In this case, calculations based on equilibrium constants is sometimes the only feasible approach to the identification of the species. A change in the ambient physicochemical conditions often disturbs the existing acid-base, redox and complexation equilibria in such a way that the species finally determined may not reflect those that originally existed in the sample.

Elemental speciation can be addressed at different levels:

- nuclear level: an isotope distribution can provide information on the environmental source or biogeochemical fractionation. No sample preparation or separation procedure affect speciation at this level.
- electronic level: the redox state of the element. Uncontrolled reduction or oxidation are common during sampling, storage and any chemical treatment of the sample.
- organometallic level: the covalent bond between the heteroatom and carbon atom, *e.g.* Bu_3Sn^+, the counter ion is unimportant.
- coordination complex level. Exchange of the metal or of the ligand is common.

3 Stability of Species

A number of studies of the stability of different organometallic species in a variety of matrices stored at different temperatures and light exposure conditions have been carried out.[1] Organometallic species are usually stable during the analytical procedure unless present at very low concentrations or subject to harsh extraction conditions. On the other hand, metal complexes with biomolecules are readily subject to deterioration.

Since complexation equilibria between complexes of metals with biomacromolecules and 'free' metals are strongly pH-dependent, the control of the pH of the mobile phase is crucial irrespective of which separation technique is used. For example, in the case of metal complexes with metallothioneins, acidic pH is responsible for the depletion of metals leading gradually to apo forms. For instance, Zn is lost from MT at pH 5, at pH 3 Cd_4 adducts of the isoform are present, at pH 2 Cu still remains attached.[2,3] Various intermediate partially metallated forms occur at various pH ranges. The buffer chosen should therefore ensure that speciation of the analyte is not altered during its passage through the separation method. In the analysis of body fluids a buffer solution corresponding to the physiological conditions (pH 7.4) is usually selected in order to avoid denaturation of the proteins during the chromatographic separation.

The native form of the proteins in the separation system must be preserved. In this context polymeric supports or silica gel matrices carrying different polymeric chains with active ends have been investigated.[4] The saline concentration of the mobile phase is an important parameter affecting the elution volume of proteins. It affects the protein conformation and hydrodynamic radius and needs to be carefully optimised.[5] Transport proteins are most influenced by the separation conditions; in the case of metalloenzymes the metal ion is strongly bound within the structure of the protein. The nonaggressive separation conditions used in SEC do not denature these complexes.

As part of an investigation into the stability of metal-protein complexes in biological fluids, different buffers (varying in molarity and pH) have been compared with H_2O as the mobile phase in SE HPLC in order to investigate differences between the Zn distribution pattern and the stoichiometry of the complex. H_2O was found to be a more suitable mobile phase than buffers with regard to contamination. A method was developed to investigate the possibility of the transfer of Zn among the proteins during SEC; casein and metallothionein were chosen as competitive Zn ligands.[6] No change in the Zn status of the protein was detected, indicating a stable protein-metal complex under the experimental conditions.[6]

4 Recovery

Extraction/preconcentration of the analytes from the matrix prior to introduction onto a chromatographic column is a common step in each analytical procedure, with the possible exception of biological fluids. Ideally, an analyte should be separated without loss or change in speciation. In practice fulfilling of these requirements is difficult since neither the quantity of a naturally existing species nor its identity are known. An extensive discussion on recovery errors during extraction and preconcentration is beyond the scope of this book and can be found elsewhere.[1,7,8]

Chromatographic recovery has been paid less attention in elemental speciation, probably because most procedures for which validation issues have been addressed were based on gas chromatography. The recovery of metal complexes from size exclusion chromatographic columns was reported to be problematic.[6] Spiking cannot be regarded as an aid to the verification of recovery since a selective enhancement of metal concentrations is possible due to the free metal-binding capacities of certain proteins.[9] Also, a depletion of metal ions added to the mobile phase has been observed in the separation of proteins with free metal-binding capacities.[9]

Non-specific binding of metal ions or proteins to the column stationary phase is common. Possible ways to limit this include a reduction in the number of nonspecific binding sites by using $NaBH_4$ or the continuous saturation of all binding sites by adding the relevant metal ion to the buffer system. Stabilisation of the elution pattern may be achieved by methylation of the binding sites.[10] The use of organic vinyl copolymers is recommended.[11] In order to prevent changes in metal binding the separation conditions must be optimised and the use of radioactively marked isotopes offers a sensitive way to do this. The complex of interest can be labelled *in vivo* for animal subjects but for humans only *in vitro* labelling is possible.

The optimum eluent should be chosen to ensure minimum competition between buffer and cytosolic ligands and for the reduction of chemical interactions between these ligands and the gel.[12] Dilute buffers are not acceptable because of adsorption of low molecular weight compounds by the column packing. The addition of metal ions to the chromatographic mobile phase has been proposed to prevent the dissociation of the intrinsic elemental species and to increase the recovery. Formation of new complexes with high binding stability may result.[9]

5 Contamination Risk

The risk of contamination at the sampling level is at least as acute as in the trace analysis of total metals and the same precautions should be applied. The contamination hazard consists not only of the addition of an exogeneous analyte element to the sample but also that the contaminant element may exchange analyte metals with species present in the sample and hence change the initial speciation.

Contamination is a problem when analysing samples from pristine areas such as organolead compounds from Greenland snow[13] or organotin in oceanic water.[14] Its primary source is the use of derivatisation reagents (Grignard reagents, $NaBEt_4$) that are very difficult to purify. The problem is more acute in the speciation analysis of mercury, since the common derivatisation reagents, $NaBH_4$ and $NaBEt_4$, contain readily detectable quantities of Hg^{2+}.

In the case of speciation analysis of metal complexes the control of contamination is more problematic since it concerns all the metal ions that may exchange with the metal coordinated by the ligand of interest. Protocols for sampling of milk and standard specifications for cleaning the material used have been discussed in detail.[15] The procedures seem likely to be suitable when extended to other biofluids. Chromatographic buffers have been cleaned by elution over Chelex 100 and subsequently checked for contamination.[5,15] A silica-based scavenger column placed in the proximity of the injection valve has been proposed in order to retain any Al and Fe originating from buffer solutions and recipients.[16] A column of C_{18} silica impregnated with Chelex 100 placed between the chromatographic pumps and the injection valve has also been proposed to minimise the exogenous contamination.[17] An ion exchange column has also been used as a scavenger to retain contaminating elements from the eluent.[18] A metal-free HPLC-ETAAS system for the separation of the proteins and of the inorganic/

organic metal species of interest allowing the quantitative recovery has been described.[16]

Platinum released from the electrodes in PAGE can apparently react with biomolecules and give rise to contamination artefacts.[19]

6 Standardisation

Standards for anthropogenic contaminants are generally readily available in good (>98%) purity. There has also been progress regarding the availability of CRMs but they usually remain too expensive to be used as calibrants.[1]

A different situation is faced by an analyst interested in naturally occurring species in biological tissues. Not only reference materials but even simple standards are unavailable for the majority of metal species of interest. Some metal-protein complexes have been identified, but the majority, especially those at lower trace and ultratrace levels have as yet remained undiscovered. To date, the application of hyphenated techniques to bioinorganic chemistry has been more exploratory (looking for new species) than confirmatory (determining an expected compound).[20] The two acute problems associated with the reliability of biochemical speciation analysis are:

(1) whether a signal produced by the detector belongs to one particular compound.
(2) the identification of this compound.[20]

Most data have been obtained by matching the retention time with available standards. However, in addition to the limited availability of standards mentioned above, this method may lead to a number of errors.[21] The use of different (complementary, orthogonal) separation mechanisms to assure the chromatographic purity of the species arriving at the detector and to minimise the risk of misidentification has been advocated.[20-23] A recent study on speciation of arsenic showed a considerable risk of signal overlap in the analysis of algal samples containing arsenosugars.[24]

The problem of artefact formation should be mentioned. Any organic species that adheres to the column can also bind inorganic species giving rise to anomalous peaks in subsequent runs.[25] The use of a new guard column for each injection and an extensive column cleanup are recommended.[25]

As discussed above molecular mass spectrometric techniques, such as ES MS or MALDI TOF MS play an increasingly important role in the identification of trace element metal and metalloid species but require multistep fractionation and purification procedures.

In many applications SEC-ICP MS has been used as a semi-quantitative technique to monitor the relative changes in analytical signals in a well defined series of samples. Quantification of signals is usually done using peak area calibration either by converting (by removing the column) the measurement system into a flow-injection mode after completing the chromatographic run,[26,27] or using a calibration graph if standards are available.[28,29,45] ICP MS instrumental instability due to the clogging of the plasma torch, sampling and skimmer orifices is common. The correction by the post-column addition of an internal standard is recommended.

7 Isotope Dilution Analysis

Isotope dilution (ID) analysis is considered to be a definitive measurement technique. In addition to the elimination of problems associated with instrumental drift, incomplete recoveries and matrix effects, it offers precision and accuracy unmatched by other techniques. ID analysis cannot overcome problems occurring during species transformation or due to contamination. The equilibration of the spike and analyte, attainable in classical ID MS by multiple sequential dissolution and evaporation-to-dryness cycles, cannot be guaranteed to be achieved for speciated ID analysis in solid samples. Consequently, the prerequisite of the ID method, that the spike is added in an identical form to the analyte is extremely difficult, if not impossible, to attain. Nevertheless, some advantages, such as inherent corrections for the loss of analyte during sample preparation, for incomplete derivatisation yield, and for intensity suppression/enhancement in the plasma are evident. In particular, ID quantification seems to be attractive in speciation analysis of complex matrices (*e.g.* gas condensates) when the different organic consituents of the sample continuously modify the conditions in the plasma and thus the sensitivity.[30] Most ID analytical applications have been carried out using ICP MS detection although a method using EI MS has been developed for speciation of organotin.[31]

Principle of Isotope Dilution Analysis

The fundamentals of ID-ICP MS for species-specific analysis have been outlined by Gallus and Heuman,[32] Hill *et al.*[33] and Huo *et al.*[34] The prerequisite for the technique is that the analyte of interest should have more than one stable isotope and that the isotope can be determined without interference by the detector. These conditions are fulfilled for most elements of interest in speciation analysis, except arsenic. Speciated ID analysis is only possible for element species well defined in their structure and composition. The species must not undergo interconversion and isotope exchange prior to separation.

The reference isotope is usually the isotope of highest natural abundance, while the spike isotope is generally one of the less abundant natural isotopes. A spike chemically identical to the species to be determined needs to be synthesised using the less abundant isotope. In addition to the skill required to prepared an isotopically labelled spike, wider use of the technique can be limited by the substantial cost of the isotopically enriched element.

The spiked sample is analysed according to the analytical procedure to give a chromatogram (electropherogram) containing peaks at a given retention time corresponding to the reference and spiked isotope. The concentration of the analyte is calculated according to the equation:

$$C_x = C_s W_s / W_x A_s - RB_s / RB_x - A_x \tag{3}$$

Where:
C_x and C_s denote the concentrations of the analytes in the sample and in the spike solution, respectively, W_s and W_x denote the weight of spike and weight of sample, respectively, B_s and B_x the abundance of spike isotope in the spike and sample, respectively and R is the ratio between reference and spike isotopes in the sample after spiking. This calculation needs to be performed for a defined peak in the chromatogram.

The stability of the spike solution should be verified by reversed isotope dilution analysis. This consists of treating the spike solution as a sample and spiking it with a fresh standard (having the natural isotopic abundance). The sample and the spike values in the equation above are interchanged.

Choice of Detector

The precision of the isotope ratio measurement offered by a quadrupole mass analyser are sufficient for the purpose of isotope dilution analysis provided that problems of mass bias correction and detector dead time are properly addressed.

The accuracy of the isotope ratio measurement is affected by the mass discrimination (mass bias) effect which leads to an apparent increase in the measured intensity for heavier isotopes. This effect needs to be corrected by analysing a reference material with a certified isotope ratio (*e.g.* NIST 981 for lead) and comparing the measured ratio with the certified value. A correction factor calculated in this way is then applied to analytical results. An alternative can be the measurement of the Tl isotope pair (added postcolumn) but a multicollector instrument or at least a time-of-flight instrument is necessary to avoid skew on the short transient signals[35,36] when the number of measured isotopes increases.

Another problem is related to the dead time, *i.e.* the time before and after each pulse during which a detector operating in the pulse-counting mode does not record. An ion impinging on the surface of the detector during the dead time is not recorded and the actual count rate is higher than the observed rate. The effect of the dead time on the count rate of an isotope can be substantial, especially at larger count rates (higher concentrations). The dead time should be corrected on a periodic basis (it changes with the age of the detector) according to the equation:

$$[N_T = N_0 / 1 - N_0 \tau] \tag{4}$$

where N_0 is the measured count rate, N_T the actual count rate and τ the detector dead time.

Examples of Speciated Isotope Dilution Analysis

Applications have included the accurate analysis of butyltins and and mercury species. The determination of dibutyltin in sediment has been carried out by ID analysis using a ^{118}Sn-enriched spike. No recovery corrections for aqueous ethylation or extraction into hexane were necessary and no rearrangement reactions were evident from the isotope ratios.[37] A mixed spike containing ^{119}Sn enriched mono-, di- and tributyltin has been prepared by direct butylation of ^{119}Sn metal and characterised by reversed isotope dilution analysis by means of natural mono-, di- and tributyltin standards. The spike characterised in this way was used for the simultaneous determination of the three butyltin compounds in sediment CRMs.[38] Isotopically labelled Me_2Hg, $MeHgCl$ and $HgCl_2$ species have been prepared and used for the determination of the relevant species in gas condensates with detection limits in the low pg range.[30]

Examples of HPLC-ICP MS include species-specific determination of Me_3Pb^+ and Et_3Pb^+ in artificial rain water[39,40] and road dust samples,[41] organotin compounds in sediments[33,42,43] and iodine[44,45] and chromium in environmental samples.[46] Post-column addition of unspecific spike has been used for the quantification of dissolved Cu and Mo species.[47] Species-unspecific spike has been used for the accurate quantification of S and metallothionein-coordinated metals (Cd, Cu and Zn).[48]

8 Interlaboratory Studies and Certified Reference Materials

Methods and dedicated instruments for speciation analysis are still under development, and typical intra- and interlaboratory standard deviations are substantial. Accuracy in most speciation measurements using hyphenated techniques is usually difficult to demonstrate owing to the multiplicity of unknown variables. The introduction of isotopically labelled species may improve this situation although there is still a lot to be done to assure the mixing of the spike with the matrix and the stability of the coordination complexes.

During the last decade, international collaborative efforts (through interlaboratory 'round-robin' exercises and certification of reference materials) have enabled a systematic study of hyphenated techniques used for the determination of chemical species *e.g.* arsenic, mercury, lead, tin and selenium in environmental matrices (water, fish or mussel tissues or sediments). These studies, extensively discussed elsewhere[1,49] led to the establishment of reference points (*e.g.* certified values in reference materials) that offer a value with which laboratories may compare their data internationally, and, hence, achieve traceability. CRMs available for speciation analysis are summarised in the chapters devoted to particular elements. Most of them concern methylmercury in biological tissues (fish, lobster and mussels) and organotin in marine biota and sediments.

Whereas the efforts of BCR, NRCC, NIST and NIES have ensured the availability of CRMs for most organometallic compounds of environmental interest, virtually none are available for metal complexes. The problem with reference materials for speciation analysis is their stability over time and the cost of preparation, especially in view of the increasing number of species of interest in a variety of matrices. Therefore the development of reference measurement procedures, equally applicable to real sample matrices, seems to be a more desirable trend in the coming years, than the development of thousands of RMs.[50]

Many speciation analyses for endogenous species have been limited to qualitative analysis (species identification) only and the problem of the reliability of quantification appears as a remote although essential goal.

References

1. P. Quevauviller, *Method performance studies for speciation analysis*, Royal Society of Chemistry, Cambridge, 1998.
2. M.P. Richards, *Meth. Enzym.*, 1991, **205**, 217.
3. R. Lobinski, H. Chassaigne and J. Szpunar, *Talanta*, 1998, **46**, 271.
4. K. Pomazal, C. Prohaska, I. Steffan, G. Reich and J.F.K. Huber, *Analyst*, 1999, **124**, 657.
5. E. Coni, A. Alimonti, A. Bocca, F. La Torre, E. Menghetti, E. Miraglia and S. Caroli, *Trace Elem. Electrolyt.*, 1996, **13**, 26.
6. B. Michalke and P. Schramel, *J. Trace Elem. Electrolyt. Health Dis.*, 1990, **4**, 163.
7. P. Quevauviller, *J. Chromatogr. A*, 1996, **750**, 25.
8. P. Quevauviller and R. Morabito, *Trends Anal. Chem.*, 2000, **19**, 86.
9. H. Reinecke and L. Dunemann, *Fresenius' J. Anal. Chem.*, 1990, **338**, 630.
10. H. Sunaga and K.T. Suzuki, *J. Liq. Chromatogr.*, 1988, **11**, 701.
11. A. Raab and P. Braetter, *J. Chromatogr. Biomed. Appl.*, 1998, **707**, 17.
12. S. Micallef, Y. Couillard, P.G.C. Campbell and A. Tessier, *Talanta*, 1992, **39**, 1073.
13. R. Lobinski, C.F. Boutron, J.P. Candelone, S. Hong, J. Szpunar Lobinska and F.C. Adams, *Anal. Chem.*, 1993, **65**, 2510.
14. H. Tao, C.R. Ramaswamy Babu Rajendran, C.R. Quetel, T. Nakazato, M. Tominaga and A. Milyazaki, *Anal. Chem.*, 1999, **71**, 4208.
15. Y. Makino and S. Nishimura, *J. Chromatogr. B*, 1992, **117**, 346.
16. P.C. d'Haese, G.F. Van Landeghem, L.V. Lamberts and M.E. De Broe, *Mikrochim. Acta*, 1995, **120**, 83.
17. A.B. Soldado Cabezuelo, E. Blanco Gonzalez and A. Sanz-Medel, *Analyst*, 1997, **122**, 573.
18. I. Leopold and B. Fricke, *Anal. Biochem.*, 1997, **252**, 277.
19. S. Lustig, D. Lampaert, K. De Cremer, J. De Kimpe, R. Cornelis and P. Schramel, *J. Anal. At. Spectrom.*, 1999, **14**, 1357 .
20. R. Lobinski, I.R. Pereiro, H. Chassaigne, A. Wasik and J. Szpunar, *J. Anal. At. Spectrom.*, 1998, **13**, 859.
21. J.M. Gonzalez LaFuente, J.M. Marchante-Gayon, M.L. Fernandez Sanchez, and A. Sanz-Medel, *Talanta*, 1999, **50**, 207.

22. J. Szpunar and R. Lobinski, *Fresenius' J. Anal. Chem.*, 1999, **363**, 550.
23. R. Lobinski and M. Potin Gautier, *Analysis.*, 1998, **26**, M21.
24. S. McSheehy, M. Marcinek, H. Chassaigne and J. Szpunar, *Anal. Chim. Acta*, 2000, **410**, 71.
25. H.M. Crews, J.R. Dean, L. Ebdon and R.C. Massey, *Analyst*, 1989, **114**, 895.
26. J. Szpunar, P. Pellerin, A. Makarov, T. Doco, P. Williams, B. Medina and R. Lobinski, *J. Anal. At. Spectrom.*, 1998, **13**, 749.
27. K. Takatera and T. Watanabe, *Anal. Sci.*, 1993, **9**, 19.
28. K.A. High, B.A. Methven, J.W. McLaren, K.W.M. Siu, J. Wang, J.F. Klaverkamp and J.S. Blais, *Fresenius' J. Anal. Chem.*, 1995, **351**, 393.
29. J. Szpunar, H. Chassaigne, O. Donard, J. Bettmer and R. Lobinski, in G. Holland and S. Tanner (Eds.), *Plasma Source Mass Spectrometry, Developments and Applications*, Royal Society of Chemistry, Cambridge, 1997.
30. J.P. Snell, I.I. Stewart, R.E. Sturgeon and W. Frech, *J. Anal. At. Spectrom.*, 2000, **15**, 1540.
31. C.G. Arnold, M. Berg, S.R. Muller, U. Dommann and R.P. Schwarzenbach, *Anal. Chem.*, 1998, **70**, 3094.
32. S.M. Gallus and K.G. Heumann, *J. Anal. At. Spectrom.*, 1996, **11**, 887.
33. S.J. Hill, L.J. Pitts and A.S. Fisher, *Trends Anal. Chem.*, 2000, **19**, 120.
34. D. Huo, H.M. Kingston and B. Larget, in J.A. Caruso, K.L. Sutton and K.L. Ackley (Eds.), *Elemental speciation. New approaches for trace element analysis*, Elsevier, Amsterdam, 2000.
35. E.M. Krupp, C. Pécheyran, H. Pinaly, M. Motelica-Heino, D. Koller, S.M.M. Young, I.B. Brenner and O.F.X. Donard, *Spectrochim. Acta*, 2001, **56B**, 1233.
36. E.M. Krupp, C. Pécheyran, S. Meffan-Main and O.F.X. Donard, *Fresenius J. Anal. Chem.*, 2001, **370**, 573.
37. J.R. Encinar, J.I. Garcia Alonso and A. Sanz-Medel, *J. Anal. At. Spectrom.*, 2000, **15**, 1233.
38. J.R. Encinar, M.I.M. Villar, V.G. Santamaria, J.I.G. Alonso and A. Sanz-Medel, *Anal. Chem.*, 2001, **73**, 3174.
39. A.A. Brown, L. Ebdon and S.J. Hill, *Anal. Chim. Acta*, 1994, **286**, 391.
40. L. Ebdon, S.J. Hill and C. Rivas, *Spectrochim. Acta*, 1998, **53B**, 289.
41. P. Quevauviller, R. Harrison, F. Adams and L. Ebdon, *Trends Anal. Chem.*, 2000, **19**, 195.
42. C. Rivas, L. Ebdon, E.H. Evans and S.J. Holl, *Appl. Organomet. Chem.*, 1996, **10**, 61.
43. L. Yang, Z. Mester and R.E. Sturgeon, *Anal. Chem.*, 2002, **74**, 2968.
44. K.G. Heumann, L. Rottmann and J. Vogl, *J. Anal. At. Spectrom.*, 1994, **9**, 1351.
45. K.G. Heumann, S.M. Gallus, G. Raedlinger and J. Vogl, *Spectrochim. Acta*, 1998, **53B**, 273.
46. H.M. Kingston, D. Huo, Y. Lu and S. Chalk, *Spectrochim. Acta*, 1998, **53B**, 299.
47. L. Rottmann and K.G. Heumann, *Fresenius' J. Anal. Chem.*, 1994, **350**, 221.

48. D. Schaumloeffel, A. Prange, G. Marx, K.G. Heumann and P. Braetter, *Anal. Bioanal. Chem.*, 2002, **372**, 155.
49. P. Quevauviller, *Fresenius' J. Anal. Chem.*, 1996, **354**, 515.
50. P. De Bievre in H. Günzler (Ed.), *Accreditation and Quality Assurance in Analytical Chemistry*, Springer, Berlin, 1996.

Part II Applications

CHAPTER 8

Multielement Analysis for Organometallic Species in the Environment

1 Introduction

In addition to their natural occurrence, metals and metalloids are also found in sewage sludge and in municipal and industrial waste deposits as a result of various household and industrial discharges. Because of microbiological activity and reducing conditions in such matrices, the metal(loid)s can be biomethylated and volatilised. Some of the species formed are very toxic so their concentrations in ambient or workplace air are of special concern and may require continuous monitoring.

Volatile metal(loid) compounds have been identified in a variety of anthropogenic gases, *e.g.* landfill gas and sewage sludge digester gas. These compounds are non-charged hydrides and/or methylated or alkylated compounds of main-group elements of groups 12 to 17. Compounds such as dimethylmercury (Me_2Hg), dimethyl selenide (Me_2Se), methyl-, butyl-, and phenyl tin, trimethyl-stibine (Me_3Sb), trimethylbismuthine (Me_3Bi), methylated arsines (Me_xAsH_y, $x + y = 3$), dimethyltelluride (Me_2Te), alkylated lead (Et_xMe_yPb, $x + y = 4$) have been identified in concentrations ranging from $ng\,m^{-3}$ to $\mu g\,m^{-3}$.[1,2] Carbonyls of Ni, Mo and W have been found in sewage gas and landfill gas.[3,4]

The variety of species of a number of different elements require analytical approaches which enable a rapid, simultaneous multielement, multiisotopic and multispecies analysis (up to 30 species within a single run). In view of the volatility of the methylated species gas chromatography is the preferred separation technique. The prerequisite for the separation of several compounds within one chromatographic run is similar physicochemical properties (boiling point within 200–300 °C) and reasonable thermal stability. An ICP quadrupole MS offers the possibility of simultaneous element-specific detection of up to eight isotopes. More can be co-determined when a TOF analyser is applied.

The metal species of interests include:

(1) native organometallic species present in the gas phase, dissolved in water or adsorbed on soil and sediments that have the valence states of a metal saturated by hydrogen atoms or alkyl groups.
(2) ionic species in natural waters, soils and sediments that can be converted (by means of a derivatisation reaction) in to a series of hydrides with similar physicochemical properties.

For higher boiling compounds, the multielement multispecies approaches are limited to the simultaneous speciation of mercury, lead and tin derivatised by means of ethylation[5] or propylation.[6,7]

2 Cryogenic Trapping Followed by Low Temperature GC-ICP MS

An analytical setup for a multielement multispecies low temperature GC speciation analysis is shown schematically in Figure 8.1.[8] Its advantage is the possibility of the simultaneous determination of different quantities of an element species from upper femtogram to microgram levels. The setup is composed of a purge vessel, a cryofocusing trap in the form of a U-tube filled with a chromatographic sorbent, and a low-temperature GC unit, usually using a *ca.* 1 m glass column connected by means of a transfer line to the ICP MS torch. The purge vessel may be replaced by a hydride generation unit. The purge vessel or the hydride generation unit are redundant in the analyses of gases and are by-passed.

The method of simultaneous generation of hydrides from trace element species in environmental samples, developed and improved over the years by the group of Hirner,[9–13] is based on reacting a water sample or a slurry of soil or sediment with $NaBH_4$ at pH 1 followed by drying the gas phase using an $MgClO_4$ trap prior to cryofocusing at $-196\,^{\circ}C$ in a U-tube trap. The methylated species/hydrides are separated by low temperature GC using a glass column allowing temperature programming from $-196\,^{\circ}C$ to $150\,^{\circ}C$. Because of the high reactivity of the compounds the use of metal-free PTFE material is recommended. The low thermal stability of the hydrides requires the desorption temperatures be kept as low as possible.

ICP MS has been successfully used as the principal detection technique, together with a quadrupole mass analyser. The need for a truly multielement analysis and high precision in the isotope ratio determination makes the use of ICP TOF MS an attractive alternative. The combination of cryotrapping GC and ICP TOF MS has been proposed for isotope ratio measurements in multi-tracer experiments and isotope dilution methodology.[14]

Packed column technology can be replaced by capillary traps and capillary column separations. The latter, preceded by cryotrapping in a capillary insert

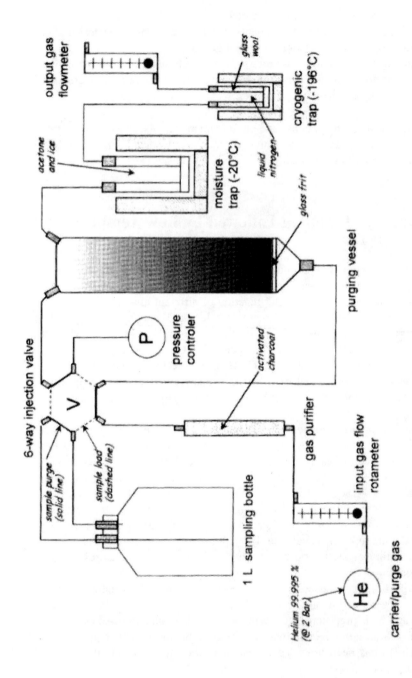

Figure 8.1 *A setup for multielement multispecies sampling of volatile metal(oid) species in natural waters by in situ purge and cryogenic trapping followed by gas chromatography and inductively coupled plasma mass spectrometry (Reprinted from Anal. Chim. Acta, 1998, 359, 227, copyright 1998, with permission from Elsevier)*

offers lower detection limits because of more concentrated bands. Consequently smaller quantities of samples can be used in order to attain similar detection limits to packed columns. Gas samples of 10–100 mL (standard atmospheres) and 500 mL (landfill gas) have been cryotrapped on 40 cm fused-silica tubing submerged in liquid nitrogen.[15] The minimum cryotrapping temperature allowed is a function of the composition of the analysed atmosphere. For example for landfill gas samples rich in methane, trapping at liquid nitrogen temperature bears the risk of blocking the trap with methane.[15] A temperature of −80 °C, at which methane is not retained, is recommended.[15]

Analysis of Air and Atmospheric Samples

Sampling remains the most critical point in the analysis of air for organometallic species. Solvent or absorbing liquids cannot be used for sampling because the interactions needed for the analyte to be caught in the absorbent risk changes to its chemical form. The use of adsorbing stationary phases suffers from irreversible adsorption, degradation during the desorption process, and artefact formation.[15] Therefore cryotrapping using a chromatographic packing followed by thermal desorption into an ICP MS, extensively developed by the Hirner's group and then Feldmann *et al.*,[1,3,4,16–19] is gradually gaining success among other groups[8,20,21] despite the disadvantages of using large amounts of liquid nitrogen and the need for a pump and a power supply. Usually a second trap is needed for cryofocusing in order to allow a narrow analyte band (for enhanced resolution) to enter the GC column.

The typical problems with preconcentration by cryotrapping are caused by water vapour (plugging when humid atmospheres are sampled) and the presence of CO_2 and CH_4 gases, especially when the headspace of anaerobic cultures is analysed. CO_2 can be vented after trapping before the analytes are transferred to the analytical column.[8] Carbon dioxide (and H_2O) can be eliminated on NaOH-filled cartridges and the condensation of CH_4 prevented by trapping in a dry-ice/acetone mixture instead of liquid nitrogen.[22] The recovery of alkyltin and organoarsenic species has exceeded 80% but only 30–75% of organic antimony has been recovered.[22] Oxygen addition to the carrier gas is required to reduce interferences originating from the presence of volatile carbon-containing species in the samples.[20]

The increasing sensitivity of ICP MS instruments allows the minimum sample volume necessary for analysis to be decreased. In this context sampling of volatile metal(loid) compounds such as hydrides, methylated and permethylated species of arsenic, antimony, and tin has been described using Tedlar bags prior to cryotrapping GC-ICP MS.[15]

The detection limits were at the 0.1–2.5 pg level but air quantities up to 100 L could be sampled which resulted in concentration detection limits down to 1 $pg\,m^{-3}$.[20]

Analysis of Natural Waters, Soils and Sediments

Although some non-polar species can be dissolved in water or adsorbed on soil and sediment, the majority of organometallic compounds found in natural waters and sediments have covalently unsaturated metal valencies. The generation of volatile hydrides *via* reaction with $NaBH_4$ represents the most convenient derivatisation method. Because of the high reactivity and reducing potential of $NaBH_4$ in acid media, free valences of ionic (or polarised) compounds are saturated with hydrogen atoms leading to inorganic hydrides (*e.g.* AsH_3, SeH_2) or

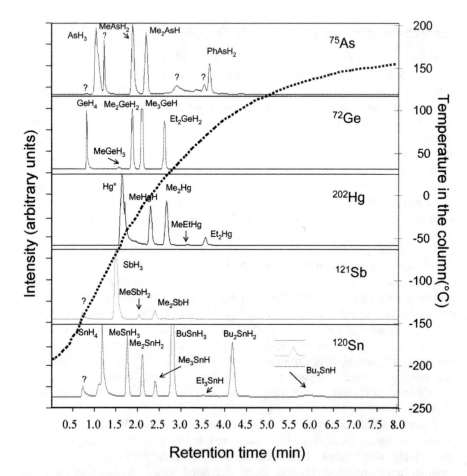

Figure 8.2 *Purge-and-trap low-temperature GC-ICP MS chromatograms for organic species of arsenic, germanium, mercury, antimony and tin derivatised as hydrides together with the temperature profile in the column during the desorption step (dotted line)*
(from Ref. 24)

mixed alkyl/hydride species (*e.g.* Me$_2$HAs). The choice of pH is a compromise between the optimum conditions for the generation of organoarsenic hydrides and those for the generation of hydrides of Bi, Ge, Pb, Sb, Se, Sn and Te. Hydride generation is known for its vulnerability to matrix in terms of reaction efficiency.[23] Also, the reducing power of NaBH$_4$ can lead to metal precipitation, which can trap and decompose the generated hydrides.

The system developed by Hirner's group enables a satisfactory separation in one run of a number of species (29 species of 12 elements have been tested) with a boiling point difference of ⩾14 °C with an absolute detection limit below 0.7 pg.[12] A modification of this system for a large volume hydride generation system (samples of 0.5–1.0 L processed at a rate of 3 h^{-1}) has been developed.[24] The hydrides are cryotrapped in a field packed column at −196 °C. The desorption is carried out in the laboratory prior to analysis by GC-ICP MS. The detection limits are in the 1–20 pg L^{-1} range.[24]

Figure 8.2 shows typical chromatograms obtained by a hydride generation purge-and-trap low temperature GC-ICP MS multielemental multispecies approach.[24] In the case of real samples the acquired chromatograms are complex, especially of elements such as Sn, As, Pb or iodine where a lot of unidentified peaks can be present. Because of the paucity of standards, identification and quantification are the main challenges in this type of analysis.

3 Identification, Calibration and Quantification

In contrast to anthropogenic organometallic contaminants, low molecular mass species are formed naturally, by hydrogenation, methylation, or other, unknown, pathways. The identification of the species formed by matching their retention time with standards is hampered because the latter are not available except as fully methylated or ethylated derivatives that have boiling points above 50–60 °C. In some cases ionic monomethyl derivatives are available that may serve for an *in situ* standard synthesis by hydride generation. Therefore a method other than retention time matching needs to be chosen for species identification.

One way is retention time matching by interpolation using correlation between the boiling point and the retention time.[12] The identification of species with boiling points from −90 to 200 °C has been achieved using a specific boiling point/retention time correlation. This is based on a 3rd order polynomial regression function built up of 26 species of Hg, Ge, Sn, Pn, As, Bi, Se and Te with a correlation coefficient of 0.99[12,23] (Figure 8.3).[23] It usually allows identification of the element hydrides, methylated species and compounds containing small (up to butyl) alkyl groups.

Standards are necessary (the correlation fails) in the case of metal(loid) organic compounds with higher molecular groups, such as monophenyl arsine, because of the additional interactions between the phenyl group and the stationary phase. Standards are also required for carbonyls such as Mo(CO)$_6$ and W(CO)$_6$.[3] A more reliable and definitive technique for the structural identification of unknown compounds is electron impact MS. Ion trap MS/MS has been proposed for the

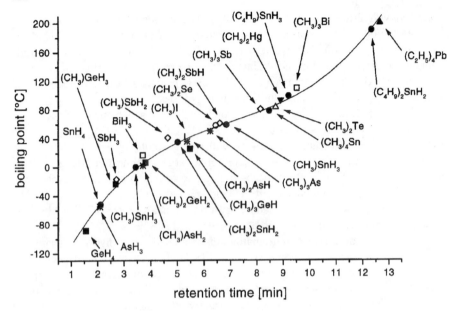

Figure 8.3 *Correlation between retention time and boiling point for organometallic species (reprinted with permission from Fresenius' J. Anal. Chem., 2000, **368**, 67, copyright 2000, Springer-Verlag)*

identification of alkyl antimony, tin and bismuth compounds in landfill and fermentation gases.[25]

In terms of quantification the main drawback is the unavailability of standard gas mixtures which could be used for calibration. As mentioned above only a few standard compounds are commercially available. The preparation of volatile metal(loid) compounds in gas samples is difficult.[15] The thermodynamic stability of gaseous standards is very low in comparison with the same species in the environment while handling of gas mixtures at very low concentrations ($pg\,L^{-1}$) is almost impossible because of adsorption-desorption effects. A method of quantification by mixing the gas sample with Rh-containing aerosol has been proposed.[18]

4 Overview of Applications

A number of different samples including urban air, landfill gas, estuary and hot-spring water, sewage sludge, and municipal and industrial waste deposits, have been analysed. In soil and aqueous samples the inorganic form of the element detected as hydride (AsH_3, GeH_4, SnH_4, SbH_3 or BiH_3) is dominant. In the case of gaseous samples the major species is usually a permethylated form such as Me_4Sn, Me_3As or Me_3Bi.

Silicium, Germanium, Tin and Lead

Trimethylsilanol has been the most abundant compound determined in waste composting gas samples, accompanied by several siloxanes.[26] Organogermanium species are not anthropogenic; partly methylated species are formed in aqueous systems.[24,27]

The source of lead in the environment is its (rapidly decreasing) use as an antiknock additive to gasoline. A typical chromatogram contains therefore of a cluster of five compounds $Me_nEt_{4-n}Pb$ that are chromatographed non-derivatised (*cf.* Chapter 12). The tetraalkylated compounds can degrade in the environment and the ionic compounds produced may form a number of species during hydride generation of which the identity is poorly defined.[12,13]

Tin shows the most complex speciation pattern. Organotins are industrially produced in large quantities as R_nSnX_{4-n}, where n = 1–3, R = methyl, butyl, *n*-octyl, or phenyl, and X, the counterion, may be chloride, maleate, acetate or laurate for application in PVC stabilizers, wood preservation additives and antifouling paints (*cf.* Chapter 9). Consequently, the full series of methyltin and butyltin hydrides are usually identified in the chromatograms as well as a number of mixed and sometimes unidentified species.[12,13]

Arsenic, Antimony and Bismuth

All three elements are known to undergo readily biomethylation in the environment. The peak of AsH_3 dominates in chromatograms of contaminated soil samples followed by the three methylarsines. $PhAsH_2$ can be identified but a number of other unidentified species are present.[12,13] Feldmann *et al.*[1,14,16,28] have analysed a variety of gas samples (domestic waste, landfill and sewage gases) discovering a number of alkylated arsines. Using a similar approach, Tseng *et al.*[24] recently reported the presence of AsH_3, $MeAsH_2$, Me_2AsH and $PhAsH_2$ in gas purged from environmental waters. AsH_3 and Me_3As have been identified in hot springs.[10]

A similar speciation pattern is observed for antimony in contaminated soil where the full range of stibines and a number of less volatile species can be detected.[12,13] Only Me_3Sb has been found in landfill gas[14] and hot springs.[10]

The pattern of bismuth is simpler. Me_3Bi is the only species in sewage[19] and in landfill gas.[14] In soil Bi^{3+} dominates with a small fraction of Me_3Bi[12] but other Bi hydrides, $BiMe_2H$ and $BiMeH_2$, have been reported in soil[13] and sediment.[9]

Selenium and Tellurium

The volatile chemical species of Se are the reduced and methylated forms. The best known are hydrogen selenide (H_2Se), methaneselenol (CH_3SeH), dimethylselenide ($(CH_3)_2Se$), dimethylselenylmethaneselenol (CH_3SeSCH_3), and dimethyl diselenide ($CH_3SeSeCH_3$). The relatively high vapour pressures of these compounds make them, in theory, significant in the biochemical cycling of Se. However, H_2Se and CH_3SeH are unstable in air and undergo rapid oxidation, the oxidised products being non-volatile. CH_3SeSCH_3 and $CH_3SeSeCH_3$ have relatively high boiling points, 132 °C and 156 °C, respectively. Therefore, the most significant compound in Se cycling is Me_2Se.

A typical chromatogram of organoselenium compounds in seawater Figure 8.4 contains three peaks corresponding to Me_2Se, Me_2SeS and Me_2Se_2. Only dimethylselenium was found in hot springs.[10] Purge-and-trap low temperature GC usually fails for compounds with a boiling point exceeding that of dimethyl diselenium. Selenium (IV) can also be ethylated, purged, cryotrapped and chromatographed; this method allows the specific detection of selenite in the sample.[29]

Knowledge of the occurrence of tellurium compounds in the environment is scarce. Me_2Te has been the sole Te species detected in landfill gas.[8]

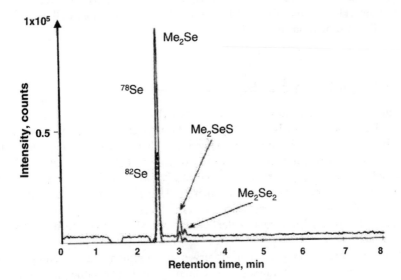

Figure 8.4 *A purge-and-trap low temperature GC-ICP MS chromatogram of organoselenium species in seawater*
(Reprinted from *Anal. Chim. Acta.*, 1998, **377**, 241, copyright 1998, with permission from Elsevier)

Other Species

Chromatograms showing methyliodide and a number of non-identified peaks corresponding to iodine compounds have been reported.[10,12] Evidence for the occurrence of $Ni(CO)_4$ in addition to $Mo(CO)_6$ and $W(CO)_6$ in fermentation gases from a municipal sewage treatment plant[3] and landfill gas[4] has been shown.

References

1. J. Feldmann and A.V. Hirner, *Int. J. Environ. Anal. Chem.*, 1995, **60**, 339.
2. A.V. Hirner, *GIT Fachz. Lab.*, 1995, **39**, 524.
3. J. Feldmann, *J. Environ. Monit.*, 1999, **1**, 33.
4. J. Feldmann and W.R. Cullen, *Environ. Sci. Technol.*, 1997, **31**, 2125.
5. M. Ceulemans and F.C. Adams, *J. Anal. At. Spectrom.*, 1996, **11**, 201.
6. L. Moens, T. de Smaele, R. Dams, P. van den Broek and P. Sandra, *Anal. Chem.*, 1997, **69**, 1604.
7. T. De Smaele, L. Moens, R. Dams, P. Sandra, J. Van der Eycken and J. Vandyck, *J. Chromatogr. A*, 1998, **793**, 99.
8. D. Amouroux, E. Tessier, C. Pecheyran and O.F.X. Donard, *Anal. Chim. Acta*, 1998, **377**, 241.
9. E.M. Krupp, R. Grümping, U.R.R. Furchtbar and A.V. Hirner, *Fresenius' J. Anal. Chem.*, 1996, **354**, 546.
10. A.V. Hirner, J. Feldmann, E. Krupp, R. Grümping, R. Goguel and W.R. Cullen, *Organic Geochemistry*, 1998, **29**, 1765.
11. A.V. Hirner, E. Krupp, F. Schulz, M. Koziol and W. Hofmeister, *J. Geochem. Explor.*, 1998, **64**, 133.
12. U.M. Grüter, J. Kresimon and A.V. Hirner, *Fresenius' J. Anal. Chem.*, 2000, **368**, 67.
13. A.V. Hirner, U.M. Grüter and J. Kresimon, *Fresenius' J. Anal. Chem.*, 2000, **368**, 263.
14. K. Hass, J. Feldmann, R. Wennrich and H.J. Stärk, *Fresenius' J. Anal. Chem.*, 2001, **370**, 587.
15. K. Haas and J. Feldmann, *Anal. Chem.*, 2000, **72**, 4205 .
16. J. Feldmann, R. Gruemping and A.V. Hirner, *Fresenius' J. Anal. Chem.*, 1994, **350**, 228.
17. J. Feldmann, T. Riechmann and A.V. Hirner, *Fresenius' J. Anal.Chem.*, 1996, **354**, 620.
18. J. Feldmann, *J. Anal. At. Spectrom.*, 1997, **12**, 1069.
19. J. Feldmann, E.M. Krupp, D. Glindemann, A.V. Hirner and W.R. Cullen, *Appl. Organomet. Chem.*, 1999, **13**, 739.
20. C. Pecheyran, C.R. Quetel, F.M. Martin and O.F.X. Donard, *Anal. Chem.*, 1998, **70**, 2639.
21. P. Andrewes, W.R. Cullen and E. Polishchuk, *Appl. Organomet. Chem.*, 1999, **13**, 659.
22. J. Feldmann, L. Naëls and K. Haas, *J. Anal. Atom. Spectrom.*, 2001, **16**, 1040.
23. U.M. Grüter, M. Hitzke, J. Kresimon and A.V. Hirner, *J. Chromatogr. A*, 2001, **938**, 225.
24. C.M. Tseng, D. Amouroux, I.D. Brindle and O.F.X. Donard, *J. Environ. Monitor.*, 2000, **2**, 603.
25. J. Feldmann, I. Koch and W.R. Cullen, *Analyst*, 1998, **123**, 815.
26. R. Grümping, D. Mikolajczak and A.V. Hirner, *Fresenius' J. Anal. Chem.*, 1988, **361**, 133.
27. Y. Kazuo Jin, Y. Shibata and M. Morita, *Anal. Chem.*, 1991, **63**, 986.

28. A.V. Hirner, J. Feldmann, R. Goguel, S. Rapsomanikis, R. Fisher and M.O. Andreae, *Appl. Organomet. Chem.*, 1994, **8**, 65.
29. M.B. de la Calle Guntinas, R. Lobinski and F.C. Adams, *J. Anal. At. Spectrom.*, 1995, **10**, 111.

CHAPTER 9

Speciation of Organotin Compounds

1 Introduction

Organotin compounds have a general formula R_nSnX_{4-n}, where n denotes an alkyl or aryl group, usually butyl-, octyl- cyclohexyl- or phenyl-, and X stands for the counterion, usually a halide, hydroxide or acetate. They have a number of industrial applications including their use as stabilisers in polymers, as catalysts in many industrial processes, such as the production of foams or silicones, and as biocides, in particular fungicides used in crop protection, antifouling paints and preservatives for timber and wood.[1]

Particular attention has been paid to the fate of the most toxic trialkyl- and triaryl derivatives released directly into the marine environment from the leaching of the tributyltin based antifouling paints used on boats and ships.[2,3] Other sources such as herbicidal use or sewage sludge discharges have been closely monitored. Ecotoxicology of organotin compounds has been comprehensively reviewed.[1] The toxicity of the compound increases with the number of R substituents, the counterion playing a marginal role, if any. Therefore, in terms of analytical chemistry the determination of total tin is irrelevant; accurate information on tributyl, triphenyl and tricyclohexyl species is needed. A very recent area of interest includes organotins leached out of polymers used for the production of hygiene products, non-wovens, and nappies.[4]

The development of species selective methodology for organotin analysis dates back to the early 80s. Since then, a large number of analytical methods have been developed that are reviewed in detail elsewhere.[5–7] The existing methodology is mature enough to be implemented in routine laboratories. Gas chromatography with MIP AED and electron ionisation MS detection are common analytical procedures but the use of ICP MS detection is expected to grow in the future. HPLC-ICP MS is a valuable independent analytical technique. However, its use in routine analysis is hampered by the need for the continuous analysis of an organic mobile phase by ICP MS and the higher risk of column damage if 'dirty' matrices are injected. The availability of reference materials offers a reliable tool for method validation. Procedures repeatedly published since the mid 90s do not seem to offer any advantage in terms of detection limits, robustness, simplicity and analysis times, with a notable exception of the isotope dilution method which allows more accurate analysis.

This chapter presents the series of methods developed, validated and comprehensively tested during the last decade in the authors' lab for speciation analysis of organotin in the most widely investigated samples: natural waters, sediments, marine biota, polymers, paints and hygiene products.

2 Analytical Techniques

A typical set of organotin species to be determined for the purpose of monitoring environmental contamination includes mono-, di- and tributyltin and mono-, di- and triphenyl tin in marine samples. Mono-, di- and trioctyl derivatives are included when hygiene products or precursor materials are analysed. The interest in the speciation of methyltin has been limited to studies of the biogeochemical cycling of trace elements in marine and terrestrial environments where they are determined as part of a multielement multispecies array in an approach discussed in detail in the previous chapter.

Separation of Organotin Compounds

The use of HPLC for the separation of organotin compounds has been scarce but the technique does allow the baseline separation of butyl and phenyltin compounds as shown in Figure 9.1.[8]

The species are separated as tropolone complexes using methanol-rich mobile phases. The use of an organic modifier negatively affects the detection limits in comparison with GC. Also, the tolerance of the HPLC columns to relatively high loads of organic matrices has not been tested enough to allow the recommendation of this approach for routine analysis. Therefore, despite the need for a prior derivatisation of organotin compounds, gas chromatography has been accepted as the separation method of choice because of its robustness, the availability of sensitive detection techniques and high sample throughput. HPLC-ICP MS, however, should be considered as a valuable independent technique to be used for method validation and the certification of reference materials.[7]

Two basic GC methods can be distinguished as a function of the derivatisation technique employed:

(1) hydride generation followed by purge and trap GC with AAS[9,10] or ICP MS detection.[11,12] Historically, this method has played an important role in the speciation of butyltins but its practical significance is now limited to studies of methyltin compounds.

(2) *in situ* ethylation and simultaneous extraction into a non-polar solvent, followed by capillary GC of the organic extract.[13] This approach is

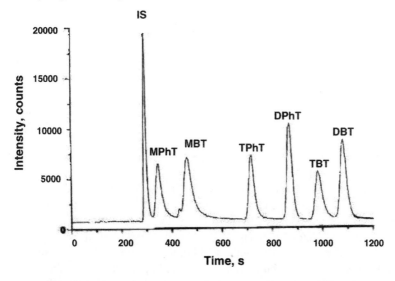

Figure 9.1. *A HPLC-ICP MS chromatogram of a mixture of butyl- and phenyltin species in water*
(Reprinted from *J. Chromatogr. A.*, 2000, **879**, 137, copyright 2000, with permission from Elsevier)

routinely used both in environmental monitoring studies (speciation of butyl- and phenyltin compounds) and in the analyses of hygiene products (butyl and octyl derivatives). The sample throughput can be increased by the use of multicapillary columns.[14]

The significance of other derivatisation techniques is marginal. A review of derivatisation methods can be found elsewhere.[15]

Detection of Organotin Compounds in Gas Chromatography

A number of techniques can be used for the detection of organotin compounds in GC.[6] They are electron impact MS, flame photometric detection (FPD), pulsed FPD, microwave emission plasma AES, and ICP MS. Despite its unquestionable role in the history of organotin speciation analysis, detection by AAS can no longer compete with the above mentioned techniques in terms of the overall performance and should be avoided.

The basic criteria for the choice of technique for organotin speciation analysis include a satisfactory detection limit and freedom from interferences. The detection limit is a function of three factors:

(1) the preconcentration factor achieved during derivatisation/extraction (ratio between the aqueous and organic phase).

(2) the injection volume (fraction of the organic extract available that is actually analyzed.

(3) the instrumental detection limit of the detector.

In a typical solvent extraction derivatisation organotin compounds are trans-ferred from 100 mL into 1 mL of organic solution which results in a preconcen-tration factor of 100. The modern electronic pressure injectors allow the injection of 5 μL of organic solvent that can be increased to 100 μL using a PTV (programmed temperature vapourisation) injector.[16] The absolute detection limits range from 4–10 pg (FPD), 1–5 pg (MS), 0.2–0.5 pg (AED), 0.1–0.4 pg (pulsed FPD) to 0.01–0.05 pg (ICP MS).

The recommended techniques for organotin speciation analysis are GC-AED and GC-ICP MS. In addition to the low detection limits their major advantages include the independence of the reponse of the compound determined, which allows quantification by internal standardisation, and a tolerance towards 'dirty' matrices. Electron impact MS performs well in the case of relatively clean extracts (natural waters, extracts of hygiene products) but the species dependent response requires the use of standard additions or isotope dilution[17] for quantifica-tion. FPD suffers from interference with sulfur, especially in the case of the analysis of sediments, detector fouling with SnO_2 and peak discrimination, especially for high boiling compounds such as triphenyltin.

Although the wider use of ICP MS as a detector is still hampered by its high acquisition and maintenance cost, the coupling of GC-ICP MS is clearly becoming the technique of choice for speciation analysis of organotins for the future.[16,18] The subfemtogram instrumental detection limits make a simple extraction from a small sample volume sufficient to obtain the desired limits of detection.[16]

3 Overview of Applications

Analysis of Water Samples

The most convenient approach is based on the method proposed by Michel and Averty[19] for butyltin compounds, which was simplified and extended to phenyltins by Ceulemans et al.[13] The method is based on the simultaneous derivatisation of quasi-ionic organotin compounds with $NaBEt_4$ and their extraction into a non-polar organic phase which is chromatographed using capillary GC. A typical chromatogram obtained by a 5 min solvent extraction using a 50:1 preconcentra-tion factor and 25 μL PTV injection with MIP AED detection (detection limit 0.1 ng L^{-1} for a 50 mL sample) is shown in Figure 9.2.[13]

The lowest detection limits reported using this approach (with ICP MS detection) are 0.01 pg L^{-1} in seawater for 1 L sample of open ocean water.[16] It should be emphasised that the extremely low detection limits reported for GC-ICP MS after an efficient preconcentration are purely academic since control of the blank below the 0.1 ng L^{-1} level is extremely difficult. Ethylation has also been used in the purge-and-trap mode but it is limited to methyl- and butyltin compounds (sensitive TBT determination is already problematic).[20]

A method becoming fashionable for the preconcentration of ethyl derivatives

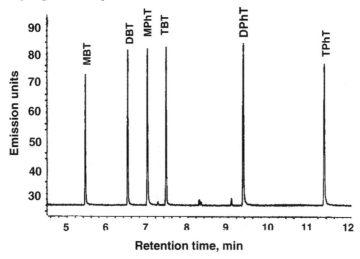

Figure 9.2. *A capillary GC-MIP AES chomatorgam for a mixture of organotin species in water*
(reprinted with permission from *Fresenius' J. Anal. Chem.*, 1993, **347**, 256, copyright 1993, Springer-Verlag)

from natural waters is solid-phase microextraction.[21-24] High preconcentration factors can be obtained leading to detection limits in the low $(0.6-20)$ pg L^{-1} range.[21] *In situ* derivatisation with NaBEt$_4$ followed by stir bar sorptive extraction (SBSE)[21-24] has allowed the reduction of DLs down to 0.1 pg L^{-1} for a 30 mL sample using GC-ICP MS.[29] The exact values for recoveries are rarely given but calculations on the basis of the data reported allow a conclusion that up to 50% of organotins from a 250 mL water sample could be sorbed on a fibre and injected from a pure water sample.[21] Caution is recommended when using SPME for routine analyses because matrix effects have not yet been sufficiently investigated.

Analysis of Sediment Samples

A considerable number of methods collaboratively evaluated and discussed elsewhere have been developed for the speciation of butyltin compounds in sediments.[25,26] The critical step is the quantitative transfer of analyte species into the aqueous phase in which they undergo *in situ* ethylation, and simultaneous extraction into a non-polar solvent, followed by capillary GC of the organic extract.

Since organotin species are not involved in mineralogical processes, the complete dissolution of the matrix is not necessary. The most convenient approach is the leaching of quasi-ionic organotin compounds with acetic or hydrochloric acid in water or a water-methanol-mixture assisted by vigorous agitation, or more often sonication.[27] Quantitative recovery of DBT and TBT is readily achieved after 1 h leaching with an acid solution but an overnight extraction is necessary for the recovery of MBT. The leaching can be considerably accelerated by using microwave-assisted extraction. Quantitative recovery of MBT, DBT and TBT has been reported after a 3 min microwave-assisted leaching.[28]

Our experience shows that acetic acid (1 + 1, *v/v*) is preferable to HCl, the latter either competing for the derivatisation reagent or inducing degradation of trialkyl(aryl) species during microwave-assisted extraction. Also, methanol should be avoided since its presence hampers the transfer of ethyl derivatives into the organic phase. The fastest aqueous-organic phase transfer is achieved with nonane but its separation from the Bu_3EtSn peak requires a careful optimisation of the GC oven programme. The use of iso-octane is a good compromise between hexane or nonane for an efficient, rapid extraction and the possibility of larger volume injection.

Figure 9.3 shows a scheme of the sample preparation procedure for speciation analysis of organotin and a representative chromatogram. The procedure has been extensively validated using different detectors: MIP AED,[14,28] QF AAS[29] and ICP MS[18] with CRMs available. In our opinion, it represents a good starting point for any speciation analysis of butyltin in sediment or soil and is rarely matched in terms of robustness, simplicity and rapidity by any other published procedure. The detection limits attained with MIP AED ($1 \, ng \, g^{-1}$) are low enough for the usual applications.

Speciation of phenyltin compounds in sediment has attracted much less attention. The preservation of the integrity of phenyltin compounds during the extraction procedure is more difficult. Pressurised liquid extraction with 0.5 M

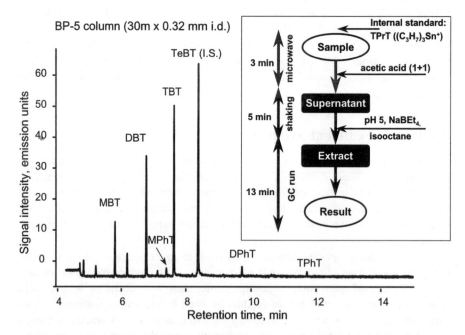

Figure 9.3. *A capillary GC-MIP AED chromatogram of organotin compounds in a sediment with a scheme of an analytical procedure shown in the inset*

acetic acid in methanol containing 0.2% (*w/v*) of tropolone has been reported as an efficient way for extraction of phenyltins from sediment samples.[8]

Analysis of Marine Biota and Seafood

A comparison study of 12 selected methods for the analysis of butyl- and phenyltin compounds in mussel samples has been reported.[30] Acidic conditions, together with the use of tropolone and a polar organic solvent are necessary for the quantitative recovery of mono- and disubstituted compounds.[30] The use of tropolone implies the use of a Grignard reagent for derivatisation that results in tedious multistep procedures.

The use of acetic acid is efficient but leads to the degradation of phenyl compounds which are widely present in biological tissues. Therefore alkaline hydrolysis with tetramethylammonium hydroxide (TMAH) is a common approach.[28] It has the advantage of totally solubilising the tissue, which is important because, in contrast to sediments, organotin compounds can be incorporated into the tissues of a living organism. TMAH hydrolysis is faster and cheaper that enzymic hydrolysis with a mixture of lipase and protease.[31] Tissue can be dissolved within 1 h of heating at 40–60 °C, the use of microwaves reducing the time necessary to a few minutes. The solution is brought to pH 5.0 and analysed by ethylation/extraction into iso-octane. Figure 9.4 shows a scheme of the analytical procedure and a typical chromatogram obtained during an analysis of fish tissue for organotin species by this method.[28] A detection limit of 2 ng/g can be obtained. The method has been validated using different detectors, MIP AED,[14,28] QF AAS[29] and ICP MS[18] with different certified reference materials.

A one step procedure based on the simultaneous extraction, ethylation with NaBEt$_4$ solution and extraction into nonane has been developed for the analysis of fish tissue.[32] Under the optimised conditions the tissue is solubilised and the derivatised organotin compounds quantitatively transfered into the organic phase. Despite the acidic reaction conditions the presence of NaBEt$_4$ assures the stability of phenyltin species which are co-determined with the butyltins.[32]

Analysis of Polymer Packagings and Hygiene Products

The unique ability of organotin stabilisers to provide transparency to poly-vinylchloride with almost no discolouration is responsible for their extensivre use in rigid (transparent) films and sheets, PVC bottles in contact with food and hygiene (non-woven) products in contact with skin. The leaching of organotins from drinking water pipes and consumer products (*e.g.* nappies, shirts, flooring) is therefore of interest.[33] The analytical protocols are based on leaching organotins with acetic acid or a diethyl dithiocarbonate (DDTC) water-ethanol mixture prior to their derivatisation with NaBEt$_4$ and GC-MIP AED or GC-MS. Organotins are also components of paints used to print packaging. Figure 9.5 shows a chromatogram of extracts from printed transparent films.

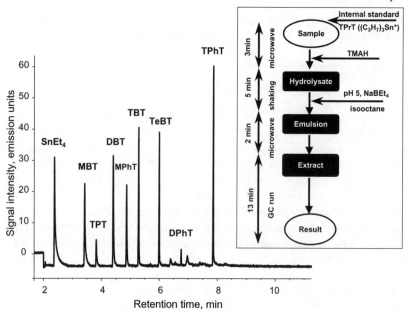

Figure 9.4. *A capillary GC-MIP AED chromatogram of organotin compounds in a biological tissue with a scheme of an analytical procedure shown in the inset*

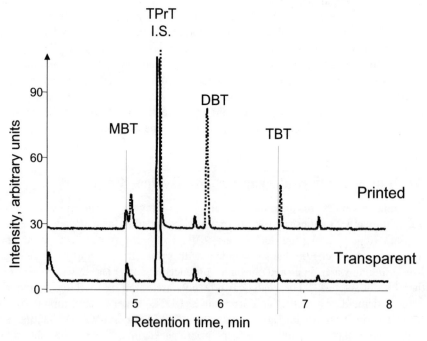

Figure 9.5. *A capillary GC-MIP AED chromatogram of plastic films*

Table 9.1. *Certified reference materials for speciation analysis of organotin*

Reference material	Matrix	MBT $(mg\,kg^{-1})$	DBT $(mg\,kg^{-1})$	TBT $(mg\,kg^{-1})$	TPhT $(mg\,kg^{-1})$
BCR 477	mussel tissue	1.50 ± 0.28	1.54 ± 0.12	2.2 ± 0.19	
NIES-11	fish tissue			1.3 ± 0.1	6.3^*
BCR 462	sediment	$(12-244)^*$	128 ± 16	70 ± 14	
PACS-1	sediment	0.28 ± 0.17	1.16 ± 0.18	1.27 ± 0.22	
PACS-2	sediment	$0.93 + 0.13$	1.09 ± 0.15	0.45 ± 0.05	
NIES-12	sediment			0.19 ± 0.03	

* indicative value
  Concentration values given as tin (PACS), cation (BCR) or chloride (NIES)

4 Method Validation

Speciation analysis of organotin compounds has been the topic of many interlaboratory studies.[34,35] The result is the availability of certified reference materials for sediments and biological tissues summarised in Table 9.1. The analysis of such material with a matrix as close as possible to that of the analysed samples should therefore be carried out.

A recent trend in method validation is the use of definitive analytical methods based on isotope dilution. Isotope dilution GC-ICP MS methods have been proposed for the accurate determination of butyltin compounds in sediments.[36–39] An alternative to ID GC-ICP MS is ID GC-MS using perdeuterated[17,40] or enriched tin-containing alkyltin compounds.[41]

References

1. K. Fent, *Crit. Rev. Toxicol.*, 1996, **26**, 1.
2. C. Alzieu, *Mar. Environ. Res.*, 1991, **32**, 7.
3. R.J. Hugget, M.A. Unger, P.F. Seligman and A.O. Valkirs, *Environ. Sci. Technol.*, 1992, **26**, 232.
4. H. Ohno, M. Suzuki, S. Nakashima, T. Aoyama and K. Mitani, *J. Food Hyg. Soc. Jap.*, 2002, **43**, 208.
5. W.M.R. Dirkx, R. Lobinski and F.C. Adams, *Anal. Chim. Acta.*, 1994, **286**, 309.
6. M. Abalos, J.M. Bayona, R. Compano, M. Granados, C. Leal and M.D. Prat, *J. Chromatogr. A*, 1997, **788**, 1.
7. M. Takeuchi, K. Mizuishi and T. Hobo, *Anal. Sci.*, 2000, **16**, 349.
8. S. Chiron, S. Roy, R. Cottier and R. Jeannot, *J. Chromatogr. A*, 2000, **879**, 137.
9. L. Randall, O.F.X. Donard and J.H. Weber, *Anal. Chim. Acta*, 1986, **30**, 184.
10. O.F.X. Donard, S. Rapsomanikis and J.H. Weber, *Anal. Chem.*, 1986, **58**, 772.

120   Chapter 9

11. E. Segovia Garcia, J.I. Garcia Alonso and A. Sanz-Medel, *J. Mass. Spectrom.*, 1997, **32**, 542.
12. K. Sato, M. Kohri and H. Okochi, *Bunseki Kagaku*, 1996, **45**, 575.
13. M. Ceulemans, R. Lobinski, W.M.R. Dirkx and F.C. Adams, *Fresenius' J. Anal. Chem.*, 1993, **347**, 256.
14. I. Rodriguez Pereiro, V.O. Schmitt and R. Lobinski, *Anal. Chem.*, 1997, **69**, 4799.
15. R. Morabito, P. Massanisso and P. Quevauviller, *Trends Anal. Chem.*, 2000, **19**, 113.
16. H. Tao, C.R. Ramaswamy Babu Rajendran, C.R. Quetel, T. Nakazato, M. Tominaga and A. Milyazaki, *Anal. Chem.*, 1999, **71**, 4208.
17. C.G. Arnold, M. Berg, S.R. Muller, U. Dommann and R.P. Schwarzenbach, *Anal. Chem.*, 1998, **70**, 3094.
18. I. Rodriguez, S. Mounicou, R. Lobinski, V. Sidelnikov, Y. Patrushev and M. Yamanaka, *Anal. Chem.*, 1999, **71**, 4534.
19. P. Michel and B. Averty, *Appl. Organomet. Chem.*, 1991, **5**, 393.
20. R. Eiden, H.F. Schoeler and M. Gastner, *J. Chromatogr. A*, 1998, **809**, 151.
21. S. Aguerre, G. Lespes, V. Desauziers and M. Potin-Gautier, *J. Anal. At. Spectrom.*, 2001, **16**, 263.
22. T. De Smaele, L. Moens, P. Sandra and R. Dams, *Mikrochim. Acta.*, 1999, **130**, 241.
23. L. Moens, T. De Smaele, R. Dams, P. Van Den Broe and P. Sandra, *Anal. Chem.*, 1997, **69**, 1604.
24. J. Vercauteren, A. De Meester, T. De Smaele, F. Vanhaecke, L. Moens, R. Dams and P. Sandra, *J. Anal. At. Spectrom.*, 2000, **15**, 651.
25. P. Quevauviller, M. Astruc, R. Morabito, F. Ariese and L. Ebdon, *Trends Anal. Chem.*, 2000, **19**, 180.
26. J. Ruiz Encinar, P. Rodriguez Gonzalez, J.I. Garcia Alonso and A. Sanz-Medel, *Anal. Chem.*, 2002, **74**, 270.
27. S. Zhang, Y.K. Chau and Q.S.Y. Chau, *Appl. Organomet. Chem.*, 1991, **5**, 431.
28. J. Szpunar, V.O. Schmitt, R. Lobinski and J.L. Monod, *J. Anal. At. Spectrom.*, 1996, **11**, 193.
29. J. Szpunar, M. Ceulemans, V.O. Schmitt, F.C. Adams and R. Lobinski, *Anal. Chim Acta.*, 1996, **332**, 225.
30. C. Pellegrino, P. Massanisso and R. Morabito, *Trends Anal. Chem.*, 2000, **19**, 97.
31. M. Ceulemans, C. Witte, R. Lobinski and F.C. Adams, *Appl. Organomet. Chem.*, 1994, **8**, 451.
32. I. Rodriguez Pereiro, V.O. Schmitt, J. Szpunar, O.F.X. Donard and R. Lobinski, *Anal. Chem.*, 1996, **68**, 4135.
33. D.S. Forsyth, R. Dabeka, W.F. Sun and K. Dalglish, *Food Addit. Contam.*, 1993, **10**, 531.
34. P. Quevauviller, *Method performance studies for speciation analysis*, The Royal Society of Chemistry, Cambridge, 1998.
35. P. Quevauviller and R. Morabito, *Trends Anal. Chem.*, 2000, **19**, 86.

36. J.R. Encinar, M.I.M. Villar, V.G. Santamaria, J.I.G. Alons and A. Sanz-Medel, *Anal. Chem.*, 2001, **73**, 3174.
37. J.R. Encinar, P. Rodriguez-Gonzalez, J.R. Fernandez, J.I.G. Alonso, S. Diez, J.M. Bayona and A. Sanz-Medel, *Anal. Chem.*, 2002, **74**, 5237.
38. J.R. Encinar, P.R. Gonzalez, J.I.G. Alonso and A. Sanz-Medel, *Anal. Chem.*, 2002, **74**, 270.
39. J.I. Garcia Alonso, J.R. Encinar, R. Gonzales and A. Sanz-Medel, *Anal. Bioanal. Chem.*, 2002, **373**, 432 .
40. T. Iwamura, K. Kadokami, D. Jin-Ya, Y. Hanad and M. Suzuki, *Bunseki Kagaku*, 1999, **48**, 555.
41. C. Bancon-Montigny, P. Maxwell, L. Yang, Z. Mester and R.E. Sturgeon, *Anal. Chem.*, 2002, **74**, 5606.

CHAPTER 10

Speciation of Organolead Compounds

1 Introduction

The widespread use of organolead compounds as antiknock additives to petrol has made lead one of the most ubiquitous metal contaminants in the environment.[1] Tetraalkyllead compounds $Me_nEt_{4-n}Pb$ ($n = 0 \div 4$) undergo a breakdown to trialkyl- and dialkyllead compounds during combustion and further in the atmosphere and all forms are scavenged from the atmosphere by rainfall.[2,3] Figure 10.1. summarises the organolead compounds emitted into the environment and the products of their degradation. Monoalkyl lead compounds are considered to be artefacts of analytical procedures and their occurrence in the environment cannot be considered as proven.

Owing to severe restrictions regarding the use of leaded petrol, the concentration levels of organolead compounds are gradually decreasing.[4] Nevertheless, a number of ionic alkyllead compounds are detected at high concentrations in road

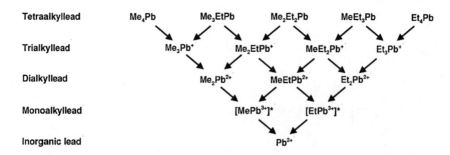

* very unstable, evidence for the presence only circumstantial

Figure 10.1 *Organolead species emitted into the environment and products of their degradation*
(from Ref. 4)

dust, road drainage and surface waters, rainwater and snow, even from very remote environments whereas tetraalkyllead compounds are still detected in urban air.[3] Organolead compounds have been detected in wine, the concentration in the vintages over the period following the rise and fall in the use of leaded petrol.[5]

The state-of-the-art of analytical methodology for organolead speciation has been reviewed.[6,7] This chapter summarises the analytical methodology for organolead speciation analysis and outlines a few procedures that we find worth recommending for this purpose.

2 Analytical techniques

Capillary GC is definitely the technique of choice for the separation of organolead species. Tetraalkyllead compounds can be chromatographed directly but the ionic products of their degradation need to be derivatised prior to GC to fully alkylated non-polar compounds. Because of the poor efficiency of hydride generation, and the impossibility of discrimination between the inorganic lead and ethyllead derivatives by ethylation, propylation and butylation have been used as derivatisation reactions. The use of Grignard reagents, PrMgCl and BuMgCl, in non-polar organic solvents (*e.g.* hexane, toluene, isooctane) has been the most widespread. Interferences during the Grignard derivatisation of organolead species have been discussed.[13] The introduction of $NaBPr_4$[8] has offered a convenient method for the derivatisation of ionic organolead compounds in the aqueous phase.[9–12] However, still needs to be validated for a wider range of samples

Historically, GC-QF AAS has played an important role in speciation analysis of organolead[6,13] but the current choice of detection technique is between MIP AED[14] and ICP MS.[12,15,16] The former technique offers remarkably low instrumental detection limits for lead (between 25 and 50 fg)[14] and competes well with ICP MS. The use of the recent generation of quadrupole analysers is likely however to reduce the ICP MS detection limits down to the sub fg level. The three coupled systems, GC-MS, GC-MIP AES and GC-ICP MS have been compared for speciation analysis of organolead.[17]

The use of HPLC is marginal, nevertheless the development of isotope dilution HPLC-ICP MS offers an important independent technique for the validation of the results obtained by the GC-based techniques.[18,19]

An emerging area of interest is the measurement of the lead isotope ratios in individual species.[20–22] As the isotopic composition of lead is not constant in nature this type of measurement can help in precise identification of Pb sources of anthropogenic contamination.

3 Overview of Applications

Tetraalkyllead Compounds in Urban Air

Tetraalkyllead species in air have been preconcentrated by cryogenic trapping and separated by gas chromatography.[23] Figure 10.2 shows a typical chromatogram of

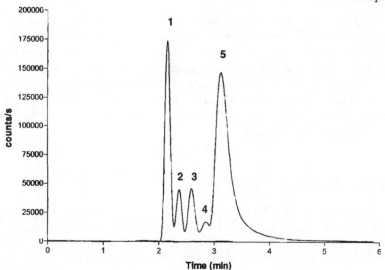

Figure 10.2 *A GC-ICP MS chromatogram of tetraalkyllead species in urban air by LT-GC-ICP MS* (1 – Me$_4$Pb, 2 – Me$_3$EtPb, 3 – Me$_2$Et$_2$Pb, 4 – MeEt$_3$Pb, 5 – Et$_4$Pb) (adapted from Ref. 24)

an urban air sample allowing the observation of the characteristic pattern of a mixture of five tetraalkyllead compounds added as antiknock additives to petrol (*cf.* Chapter 12).[24] The presence of this pattern is an unambiguous tracer of the leaded petrol origin of organolead compounds in the atmosphere.

Ionic Alkyllead in Aqueous Samples

The poor efficiency of the formation of hydrides and the impossibility of discriminating between the ethyl derivatives of ethyllead and inorganic lead practically eliminate purge-and-trap preconcentration methods unless a single, particular organolead species, such as trimethyllead is targeted. A typical procedure allowing an ultrasensitive speciation (50 pg L^{-1}) is based on the extraction of ionic alkyllead species as their complexes with diethyldithiocarbamate (DDTC) into an organic phase followed by propylation with a Grignard reagent.[14,25] The excess of inorganic lead makes its complexation with EDTA necessary prior to extraction. Preconcentration can be achieved by evaporation of the DDTC extract or large volume injection.[26] Figure 10.3a shows a scheme of a typical procedure and Figure 10.3b a typical chromatogram obtained for speciation analysis of ionic alkyllead in water samples.[14]

The above procedure has been extensively applied to the analysis of surface waters, tap water, rain water,[14,27,28] snow and polar ice,[29] and wine.[30] It may be simplified by derivatisation using NaBPr$_4$ in the aqueous phase.[8–11] Alternatively,

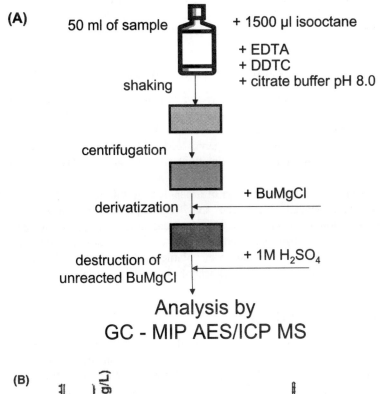

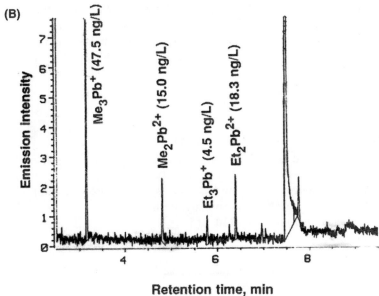

Figure 10.3 *Analysis of a water sample for organolead species: (a) a scheme of the typical procedure and (b) a typical chromatogram of a rainwater sample*

DDTC complexes of Me_3Pb^+, Me_2Pb^{2+}, Et_3Pb^+ and Et_2Pb^{2+} in rain water have been sorbed on C_{60} (fullerene) beads and eluted with a solution of $NaBPr_4$ in hexane prior to GC-TOF ICP MS.[17] A detection limit of 3 pg L^{-1} was reported.[17]

Ionic Alkyllead in Road Dust, Sediment and Biological Materials

The principle of the analytical procedure remains the same as in the case of aqueous samples. Organolead compounds are recovered from the road dust or sediment matrix by ultrasonic extraction with DDTC in the presence of a large excess of EDTA to mask the inorganic lead.[12,16,31] An exhaustive overview of methods for the determination of trimethyllead in urban dust has been published by Quevauvillier *et al.*[32]

Interest in the speciation analysis of organolead in biological samples has been much less pronounced. The same analytical approach, extraction as DDTC complexes followed by Grignard derivatisation, has been applied to speciation analysis of organolead in fish[33] and in blood.[34] A micro purge-and-trap apparatus has been developed for the isolation of volatile tetraalkyllead and butylated ionic alkyllead compounds from crude extracts of biological materials.[35]

4 Method Validation

The availability of CRMs for organolead speciation analysis is limited but corresponds to the demand. RM 604, trimethyllead in artificial rain water with a reference value of 55.2 ± 3.5 ng kg^{-1}, is suitable for the verification of calibration. BCR CRM 605, trimethyllead in urban dust with a certified value of 7.9 ± 1.2 μg kg^{-1}, serves well for testing the yield of extraction and derivatisation procedures as well as the effect of a large excess of inorganic lead on artefact formation.

References

1. J.O. Nriagu, *Sci. Tot. Environ.*, 1990, **92**, 13.
2. C.N. Hewitt and R.M. Harrison, in P.J. Craig (Ed.), *Organometallic Compounds in the Environment*, Longmans, London, 1986.
3. R.J.A. van Cleuvenbergen and F.C. Adams, in O. Hutzinger (Ed.), *Handbook of Environmental Chemistry,* Springer, Berlin, 1990.
4. R. Lobinski, *Analyst*, 1995, **120**, 615.
5. R. Lobinski, C. Witte, F. Adams, J.C. Cabanis and C. Boutron, *Nature*, 1994, **370**, 24.
6. R. Lobinski, W.M.R. Dirkx, J. Szpunar-Lobinska and F.C. Adams, *Anal. Chim. Acta.*, 1994, **286**, 381.
7. K. Pyrzynska, *Mikrochim. Acta.*, 1996, **122**, 279.
8. T. De Smaele, L. Moens, R. Dams, P. Sandra, J. Van der Eycken and J. Vandyck, *J. Chromatogr, A.*, 1998, **793**, 99.
9. M. Heisterkamp and F.C. Adams, *J. Anal. At. Spectrom.*, 1999, **14**, 1307.

10. P. Schubert, E. Rosenberg and M. Grasserbauer, *Fresenius' J. Anal. Chem.*, 2000, **366**, 356.
11. K. Bergmann and B. Neidhart, *J. Sep. Sci.*, 2001, **24**, 221.
12. M. Heisterkamp and F.C. Adams, *Fresenius' J. Anal. Chem.*, 2001, **370**, 597.
13. R.M. Harrison, C.N. Hewitt and S.J. De Mora, *Trends Anal. Chem.*, 1985, **4**, 8.
14. R. Lobinski and F.C. Adams, *Anal. Chim. Acta.*, 1992, **262**, 285.
15. M. Heisterkamp, T. De Smaele, J.-P. Candelone, L. Moens, R. Dams and F.C. Adams, *J. Anal. At. Spectrom.*, 1997, **12**, 1077.
16. I.A. Leal-Granadillo, J.I.G. Alonso and A. Sanz-Medel, *Anal. Chim. Acta*, 2000, **423**, 21.
17. J.R. Baena, M. Gallego, M. Valcarcel, J. Leenaers and F.C. Adams, *Anal. Chem.*, 2001, **73**, 3927.
18. S.J. Hill, L.J. Pitts and A.S. Fisher, *Trends Anal. Chem.*, 2000, **19**, 120.
19. L. Ebdon, S.J. Hill and C. Rivas, *Spectrochim. Acta*, 1998, **53B**, 289.
20. J. Ruiz Encinar, I.L. Granadillo, J.I. Gracia-Alonso and A. Sanz-Medel, *J. Anal. At. Spectrom.*, 2001, **16**, 475.
21. E.M. Krupp, C. Pécheyran, S. Meffan-Main and O.F.X. Donard, *Fresenius' J. Anal. Chem.*, 2001, **370**, 573.
22. E.M. Krupp, C. Pécheyran, H. Pinaly, M. Motelica-Heino, D. Koller, S.M.M. Young, I.B. Brenner and O.F.X. Donard, *Spectrochim. Acta*, 2001, **56B**, 1233.
23. C. Pecheyran, C.R. Quetel, F.M. Martin and O.F.X. Donard, *Anal. Chem.*, 1998, **70**, 2639.
24. O.F.X. Donard and R. Lobinski, *J. Anal. At. Spectrom.*, 1996, **15**, 181.
25. M. Radojevic, A. Allen, S. Rapsomanikis and R.M. Harrison, *Anal. Chem.*, 1986, **58**, 658.
26. D. Chakraborti, W.R.A. De Jonghe, W.E. Van Mol, R.J.A. Van Cleuvenbergen and F.C. Adams, *Anal. Chem.*, 1984, **56**, 2692.
27. R. Lobinski and F.C. Adams, *J. Anal. At. Spectrom.*, 1992, **7**, 987.
28. M. Paneli, E. Rosenberg, M. Grasserbauer, M. Ceulemans and F. Adams, *Fresenius' J. Anal. Chem.*, 1997, **357**, 756.
29. R. Lobinski, C.F. Boutron, J.P. Candelone, S. Hong, J. Szpunar-Lobinska and F.C. Adams, *Anal. Chem.*, 1993, **65**, 2510.
30. R. Lobinski, J. Szpunar-Lobinska, F.C. Adams, P.L. Teissedre and J.C. Cabanis, *J. Assoc. Off. Anal. Chem.*, 1993, **76**, 1262.
31. C. Witte, R. Lobinski and F.C. Adams, *Appl. Organomet. Chem.*, 1994, **8**, 621.
32. P. Quevauviller, R. Harrison, F. Adams and L. Ebdon, *Trends Anal. Chem.*, 2000, **19**, 195.
33. Y.K. Chau, P.T.S. Wong, G.A. Bengert and J.L. Dunn, *Anal. Chem.*, 1984, **56**, 271.
34. O. Nygren and C.A. Nilsson, *J. Anal. At. Spectrom.*, 1987, **2**, 805.
35. D.S. Forsyth, *Sci. Tot. Environ.*, 1989, **89**, 291.

CHAPTER 11

Speciation of Organomercury Compounds

1 Introduction

The interest in speciation of mercury is due to the enhanced toxicity of some of its specific chemical forms. Methylmercury (MeHg), the most common organo-mercury compound, is of special concern because of its ease in penetrating biological membranes, high stability and long residence time in tissues. The ability of methylmercury to accumulate in fish tissues causes mercury poisoning observed in marine life and is responsible for a risk associated with seafood consumption by man.

Mercury and organic forms of mercury are included in the black list of compounds to be monitored in the framework of national and international regulations. The development of analytical techniques for MeHg has focused primarily on seafood analysis following the poisoning accident at Minamata (Japan).[1] Interest has also been growing in biogeochemical pathways of mercury in the environment which has spurred the development of methods for the determination of trace and ultratrace levels of MeHg in water and sediment.

The interest in mercury speciation has resulted in an impressive number of papers published and still being published. The originality of most of this research is dubious, and the figures of merit are seldom improved in comparison with the earlier publications. The present analytical chemistry of mercury species is based on the methods developed by Bloom[2] in the late 80s which have undergone only cosmetic changes since then. The introduction of stable isotopes to speciation analysis of mercury by Hintelmann *et al.*[3] has improved quality control and allowed the detection of a number of sources of errors, such as the spurious formation of methylmercury during some extraction procedures.

In terms of method development, the speciation analysis of mercury has reached its maturity. Owing to the number of CRMs available analytical procedures for the analysis of sediment, fish, mussel or hair samples have been extensively validated and can be applied in routine analysis. Speciation analysis of mercury is also an area where industrial prototypes of dedicated instruments have been developed.[4]

This chapter discusses selected analytical techniques for speciation of mercury in environmental matrices: water, biota and sediment; and in clinical samples including blood, hair and soft tissues. Because of their volatility mercury species are often co-determined in multielement, multispecies arrays (*cf.* Chapter 8). Mercury speciation in petroleum matrices is discussed separately in Chapter 12.

2 Analytical Techniques

Because of the convenient conversion of mercury species into volatile compounds, gas chromatography has been the dominant separation technique prior to mercury specific detection. The high volatility and the limited number of species (usually $Hg°$, Me_2Hg, MeEtHg and Et_2Hg) allow the use of low temperature GC for the separation. A number of detectors are available; QF AAS, QF AFS, MIP AES and ICP MS have been shown to offer excellent detection limits (<50 fg). On the other hand, the sensitivity of GC-MS has been reported to be poor (30–50 pg). AAS and AFS require mercury to be introduced into a quartz furnace in the atomic form which raises the need for the post-column pyrolysis of organometallic compounds.

Purge-and-Trap Low Temperature GC Based Methods

Bloom's method (setup shown in Figure 11.1a) relies on the derivatisation of mercury species with $NaBEt_4$ at pH 4.9, followed by purging and collection of volatile organomercury compounds (Me_2Hg, MeEtHg, and Et_2Hg) on a Carbotrap packed column.[2] The preconcentrated compounds are released by heating the column and collected on a cryogenic Chromosorb packed column (80 cm × 4 mm), held at −196 °C. For analysis, the GC column is heated in a cylindrical oven at 180 °C and the outflowing carrier gas is passed *via* a 900 °C thermal decomposition tube into a cold vapour atomic fluorescence detector.[5] A typical chromatogram is shown in Figure 11.1b.

Capillary GC-Based Methods

Despite their inherent simplicity low temperature GC setups are not commercially available and the variations in design among laboratories make standardisation of methods based on this principle impossible. Speciation of mercury can be carried out using commercial purge-and-trap injectors and conventional capillary GC.[6] Mercury species in a 10–50 mL sample are derivatized with $NaBEt_4$, purged and cryotrapped. Water usually needs to be removed using a water trap cooled to −15 °C to avoid clogging the cryotrap.

An attractive alternative seems to be the use of multicapillary columns that allow the separation of mercury species isothermally and at ambient temperature thus eliminating the need for a chromatographic oven.[7]

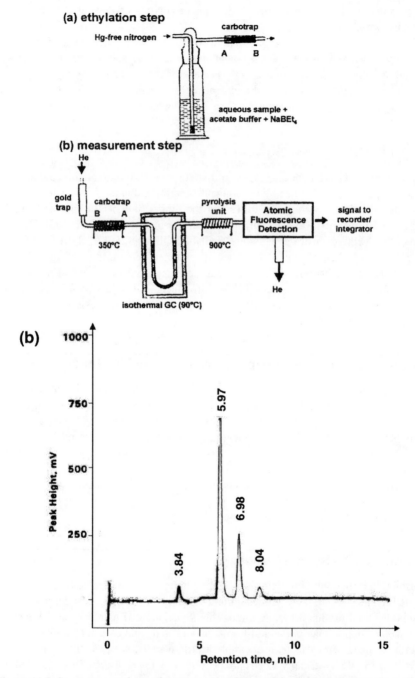

Figure 11.1 *Speciation analysis of mercury: a) scheme of the analytical setup; b) typical chromatogram obtained for a fish tissue extract using atomic fluorescence detector*
(Reprinted with permission from *Can. J. Fish. Aquat. Sci.*, 1989, **46**, 1131)

HPLC-Based Methods

Many HPLC methods have been developed for the speciation of mercury[8,9] but they cannot compete in terms of sensitivity with gas chromatography. HPLC is, however, a valuable independent analytical technique that allows control of the formation of artefacts during derivatisation found in GC. A post-column cold vapour generation unit is essential in order to obtain detection limits allowing the analysis of real-life samples (*cf.* Chapter 2).

A typical HPLC procedure includes a preconcentration of mercury species as pyrrolidinedithiocarbamate complexes on a C_{18} column followed by their gradient elution with an organic solvent. The column effluent is exposed to UV light and merged with an $NaBH_4$ solution. Hg^0 is separated in a gas-liquid separator and swept through a concentrated H_2SO_4 trap at $-20\,°C$ (to remove water and acetonitrile) to the cold vapour AFS.[10,11] Difficulties during the development of HPLC-ICP MS methods without post-column volatilisation have been discussed.[12]

3 Overview of Applications

Bloom's method can be directly applied in the analysis of both freshwater and seawater, even if they are relatively rich in dissolved organic matter.[2] The detection limit reported for monomethylmercury and dimethylmercury is $0.003\ ng\,L^{-1}$ of Hg.[2] The application of the method to solid matrices requires the transfer of organomercury species into the aqueous phase. An aliquot of the extract is considerably diluted with water (often $1 + 199$) and analysed as a water sample.

Analysis of Biological Tissues

Most sample preparation procedures prior to methylmercury determination are based on Westöö's procedure.[13] The latter includes:

(1) breaking of the Hg-S bond in the presence of an excess of halide in an acidic medium.
(2) selective extraction of organomercury into toluene.
(3) purification of the extract by back-extraction into cysteine.

The procedure, originally developed for a poorly selective detector (ECD), is too tedious and time consuming to be applied when a mercury specific detector is available. The present protocols include the leaching of methylmercury from acidic media in the presence of halide, prior to separation of methylmercury by distillation or a direct purge-and-trap procedure.

The protocol successfully used in our laboratory is based on the solubilisation of a tissue in strongly alkaline media, such as tetramethylammonium hydroxide. An aliquot of the solution in diluted with water and analysed, after *in situ* derivatisation using NaBEt$_4$, by solvent extraction capillary GC-MIP AES/ICP MS[14] or by purge-and-trap GC-AAS[15] and GC-ICP MS.[16] The tissue solubilisation procedure can be considerably accelerated by carrying it out in a microwave field.[17,18] The concentrations of interest (above 1 $\mu g\,g^{-1}$) are sufficiently high to allow the use of any of the above detectors.

Alkaline digestion is recommended as the method of choice for the recovery of methylmercury from hair samples.[19]

The speciation analysis of mercury in seafood has been reviewed.[20]

Analysis of Sediments

In contrast to biological samples, sediments contain a large excess of inorganic mercury in comparison with methylmercury. Six alternative procedures have been evaluated for the extraction of methylmercury from sediments.[21] The most satisfactory procedure in terms of quantitative recoveries and low artifact formation involves leaching a sediment sample with KBr/CuSO$_4$ solution in 5% H$_2$SO$_4$. The mixture is extracted with CH$_2$Cl$_2$ and the extract mixed with an excess of water and purged with N$_2$ to remove the organic solvent. Methylmercury remains in the aqueous phase.[21] The analysis is performed by aqueous distillation followed by aqueous-phase ethylation, purging onto a Carbotrap, thermal desorption and isothermal GC with cold-vapour atomic fluorescence detection (CV-AFD).[21]

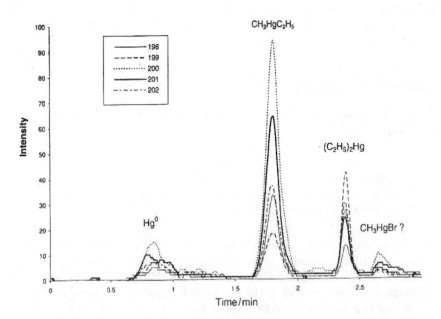

Figure 11.2 *A multiisotopic chromatogram of mercury species in a sediment extract; individual lines refer to mercury isotopes ^{198}Hg–^{202}Hg* (from Ref. 23)

In our laboratory good results have been obtained by leaching of sediments with HNO_3, the time of extraction being dramatically shortened (to *ca.* 3 min) by using a microwave field. Methylmercury in the extract was determined after derivatisation with $NaBEt_4$, either by purge-and-trap low temperature GC-AAS,[17,22] or by extraction followed by capillary GC-MIP AED.[14] The detection limit is 0.5 ng g^{-1} for a 1 g sample intake.

A multiisotope chromatogram for mercury species from a sediment sample is shown in Figure 11.2.[23]

4 Sources of Error and Method Validation

The source of error which has attracted most attention is related to the artifactual formation of methylmercury during steam distillation. The latter has long been the principal extraction technique in the determination of methylmercury in environmental matrices. It is usually based on the addition of H_2SO_4 and KCl at a temperature of 145 °C with verification of the distillation recovery by standard additions.[24] The formation of monomethylmercury artifacts during the extraction of monomethylmercury (MMHg) from water and sediments has been extensively studied.[21,24–26] The methylation of Hg(II) spikes ranges from 0.001% for rainwater and oligotrophic lake water to 0.08% for humic matter rich water. The methylation of Hg(II) spikes in sediments ranges from 0.005–0.1% with the highest conversion values observed with wetland peat. These values can be considered negligible in all cases except the analysis of sediments where the excess of Hg^{2+} over $MeHg^+$ is typically very high (>500).

Method performance evaluation for methylmercury determination in fish and sediment has been discussed in detail.[27] A number of certified reference materials from several agencies (listed in Table 11.1) are available for a variety of matrices. Speciated isotope dilution GC-ICP MS has been developed for the accurate determination of methylmercury.[28]

Table 11.1 *Certified reference materials for mercury speciation*

Reference material	Matrix	Hg mg kg^{-1}	CH_3Hg^+ mg kg^{-1} (as Hg)
BCR 463	Tuna fish	2.85 ± 0.16	3.04 ± 0.16
TORT-2 (NRCC)	Lobster hepatopancreas	0.27 ± 0.06	0.152 ± 0.013
SRM 1566b	Oyster tissue	0.0371 ± 0.0013	0.0132 ± 0.0007
DORM-1	Dogfish muscle		0.73 ± 0.06
DORM-2	Dogfish muscle		4.47 ± 0.32
DOLT-2	Dogfish liver		0.693 ± 0.053
LUTS-1	Defatted lobster hepatopancreas		0.063 ± 0.004

References

1. M. Harada, *Crit. Rev. Toxicol.*, 1995, **25**, 1.
2. N. Bloom, *Can. J. Fish Aquat. Sci.*, 1989, **46**, 1131.
3. H. Hintelmann, R.D. Evans and J.Y. Villeneuve, *J. Anal. At. Spectrom.*, 1995, **10**, 619.
4. R. Feldhaus, W. Buscher, E. Kleine-Benne and P. Quevauviller, *Trends Anal. Chem.*, 2002, **21**, 356.
5. N. Bloom and W.F. Fitzgerald, *Anal. Chim. Acta*, 1988, **208**, 151.
6. M. Ceulemans and F.C. Adams, *J. Anal. At. Spectrom.*, 1996, **11**, 201.
7. A. Wasik, I.R. Pereiro, C. Dietz, J. Szpunar and R. Lobinski, *Anal. Commun.*, 1998, **35**, 331.
8. R. Halko and M. Hutta, *Chem. Listy*, 2000, **94**, 292.
9. C.F. Harrington, *Trends Anal. Chem.*, 2000, **19**, 167.
10. R. Falter and H.F. Schoeler, *J. Chromatogr. A*, 1994, **675**, 253.
11. R. Falter and G. Ilgen, *Fresenius' J. Anal. Chem.*, 1997, **358**, 407.
12. C.F. Harrington and T. Catterick, *J. Anal. At. Spectrom.*, 1997, **12**, 1053.
13. G. Westöö, *Acta Chem. Scand.*, 1967, **20**, 1790.
14. I. Rodriguez Pereiro, A. Wasik and R. Lobinski, *J. Chromatogr. A*, 1998, **795**, 359.
15. C.M. Tseng, A. de Diego, F.M. Martin, D. Amouroux and O.F.X. Donard, *J. Anal. At. Spectrom.*, 1997, **12**, 743.
16. S. Slaets, F. Adams, I. Rodriguez and R. Lobinski, *J. Anal. At. Spectrom.*, 1999, **14**, 851.
17. C.M. Tseng, A. deDiego, F.M. Martin and O.F.X. Donard, *J. Anal. At. Spectrom.*, 1997, **12**, 629.
18. I. Rodriguez Pereiro, A. Wasik and R. Lobinski, *J. Anal. At. Spectrom.*, 1998, **13**, 743.
19. J. Yoshinaga, M. Morita and K. Okamoto, *Fresenius' J. Anal. Chem.*, 1997, **357**, 279.
20. A.M. Carro and M.C. Mejuto, *J. Chromatogr. A*, 2000, **882**, 283.
21. N.S. Bloom, J.A. Colman and L. Barber, *Fresenius' J. Anal. Chem.*, 1997, **358**, 371.
22. C.M. Tseng, A. de Diego, F.M. Martin, and O.F.X. Donard, *J. Anal. At. Spectrom.*, 1997, **12**, 629.
23. L. Lambertsson, E. Lundberg, M. Nilsson and W. Frech, *J. Anal. At. Spectrom.*, 2001, **16**, 1296.
24. H. Hintelmann, R. Falter, G. Ilgen and R.D. Evans, *Fresenius' J. Anal. Chem.*, 1997, **358**, 363.
25. H. Hintelmann, *Chemosphere*, 1999, **39**, 1093.
26. C.R. Hammerschmidt and W.F. Fitzgerald, *Anal. Chem.*, 2001, **73**, 5930.
27. P. Quevauviller, M. Filippelli and M. Horvat, *Trends Anal. Chem.*, 2000, **19**, 157.
28. R.C.R. Martin-Doimeadios, E. Krupp, D. Amouroux and O.F.X. Donard, *Anal. Chem.*, 2002, **74**, 2505.

CHAPTER 12

Metal Speciation in Petroleum-Related Samples

1 Introduction

Petroleum-related samples include natural gas and gas condensates, shale oil and products of their processing, *e.g.* petrol. The metals of interest strongly depend on the sample. Shale oils may contain considerable amounts of vanadium, nickel and manganese in the form of relatively volatile metalloporphyrins which behave differently during the distillation and refining.[1] The presence of different species of arsenic and mercury in natural gas and gas condensates raises concerns about the efficiency of the removal processes for these environmentally sensitive substances.[2] Mercury has been frequently evoked in the context of catalyst poisoning and corrosion of aluminum alloys in steam-cracker cold boxes.[3] Organometallic compounds can be more toxic than inorganic forms and negatively affect the health and safety of consumers and operating personnel. The monitoring of organolead compounds in gasoline is important from an environmental point of view, origin determination and the elucidation of arson cases. Replacing alkyllead species by organomanganese derivatives has prompted the need for methods for their species specific analysis.[4,5]

Organometallic species of concern in the petroleum industry are either volatile or can be readily converted into such by means of derivatisation. Gas chromatography has been mostly used for their separation although LC does seem to have a potential for the separation of metalloporphyrin species.[6,7] In terms of detection MIP AES and ICP MS have been successfully used. For sub-ppm detection ICP MS becomes, however, the only choice owing to its higher tolerance to complex matrices.

2 Mercury in Natural Gas and Gas Condensates

Mercury in gas condensates occurs in a wide concentration range (10–3000 $\mu g\,kg^{-1}$).[3] The prerequisite of risk assessment is the accurate measurement of its concentration, its distribution among the various distillation fractions (*e.g.* naptha, kerosene, diesel oil and the residue) and its distribution among the various chemical forms. Whereas the distribution of mercury in crude petroleum and gas condensates is well known, the forms of mercury compounds present are uncertain.[8] Evidence is growing that mercury in gas condensates is present in a variety of physical and chemical species that feature different solubilities, volatilities, toxicities and chemical reactivities.[8–10]

In contrast to natural gas that contains principally mercury in metallic form (Hg°), a variety of mercury species can be encountered in gas condensates. The presence and proportions of these species are highly dependent upon the source, the stage of production, the sampling technique used, storage of the sample and age of the sample. The majority of Hg in gas condensates seems to be associated with the particulate fraction.[10] Most dissolved mercury (>60%) is contained in the naphtha fractions (36°–170 °C).[3]

Inorganic mercury may be present in condensate in metallic (elemental, Hg°) or oxidised forms. The latter may be insoluble or colloid dispersed (HgS, Hg_2Cl_2) or soluble ($HgCl_2$). Organic compounds include organometallic compounds (with a Hg-C covalent bond) which can be fully alkylated (dialkyl mercury species) or partly alkylated (monoalkyl mercury species).

Soluble inorganic oxidised mercury is usually referred to as $HgCl_2$ (b.p. 276 °C) which is an order of magnitude more soluble than Hg° in gas condensates. Hydroxy, oxychloride and weak Hg(II) complexes with other ligands can be present as well. Insoluble mercury occurs mostly as HgS (sublimes at 584 °C) that forms a colloidal suspension in gas condensates. Particulate Hg has been found to reside in a relatively narrow particle size range, 1–10 μm and can be readily removed by filtration. The particulate fraction may also contain Hg_2Cl_2 (m.p. 400 °C) produced by reaction of Hg° and Hg(II) in the condensate upon storage.

Dialkyl mercury compounds include a range of species with different boiling points: Me_2Hg (96 °C), Et_2Hg (159 °C), i-Pr_2Hg (170 °C), Pr_2Hg (190 °C) and Bu_2Hg (206 °C).[3] Mixed alkyl species, such as MeEtHg exist, but their physicochemical properties are poorly known. The highest boiling species present seems to be Ph_2Hg (m.p. 121 °C).

Sampling and Storage

Sampling of gas condensates for mercury speciation analysis is a critical step in the analytical procedure and is potentially a source of considerable errors. Significant losses of mercury and changes in its speciation may occur by the reaction of species with the atmosphere and sample container material, and by reactions between the species themselves and between the species and other gas condensate consituents. Therefore, a gas condensate sample should be taken directly from the feed liquid prior to contact with air and water. Analysis should be carried out on-site, or, if that is not possible, the sample should be transferred into a glass or Teflon-lined container without a headspace and the time between sampling and analysis should be reduced to minimum.

Adsorption and/or reactions with the container wall are the principal reasons for the loss of mercury and changes in its speciation. Stability during storage has been investigated in detail by Bloom.[10] Metal and polyethylene (PE) containers have been found unsuitable for the collection and storage of Hg in petroleum due to either loss of Hg, or species interconversion.

Chemical reactions between $HgCl_2$ and Ph_2Hg can and between $MeHgCl$ and Ph_2Hg take place.[11] Another important reaction leading to analytical errors is the formation of Hg(I) compounds from the dissolved $Hg°$ and $HgCl_2$. Monoalkyl mercury species are stable, in contrast to $HgCl_2$, even in the presence of $Hg°$.[12]

Analytical Techniques

Analytical techniques for the speciation of mercury in natural gas and gas condensates have been comprehensively reviewed.[8,13] The risk of artefact signals is significant even with mercury-'specific' detectors such as AAS or AFS. To date, the GC-ICP MS coupling has been the only one to produce species-specific information on the whole range of dialkylmercury species in gas condensate samples at the $\mu g\,L^{-1}$ and sub-$\mu g\,L^{-1}$ levels.[14,15] In the simplest approach a gas condensate is injected onto a capillary column using HBr-containing helium as the mobile phase.[14] The non-polar species elute intact and the polar ones are derivatised to bromides and co-chromatographed. A chromatogram of a gas condensate obtained by this method is shown in Figure 12.1.

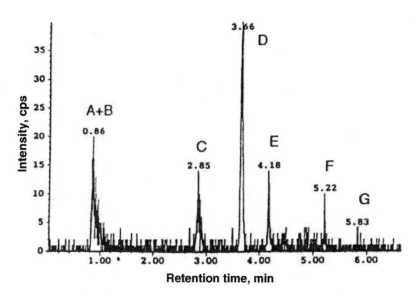

Figure 12.1 *Organomercury speciation analysis, in a gas condensate sample by GC-ICP MS: A - $Hg°$, B - $HgCl_2$, C - dimethylmercury, D - methylethylmercury, E - diethylmercury, F - methylmercury chloride, G - ethylmercury chloride (from Ref. 14).*

This method allows the determination of all the mercury species in a single run, with detection limits that are strongly species-dependent. An inconvenience remains the use of corrosive HBr in the mobile phase and the risk of species transformation in its presence at elevated temperatures. An alternative is a method using standard capillary GC conditions (He mobile phase and oven temperature gradient programming).[15] A method has been developed in which a gas condensate sample is divided in two aliquots. One is analysed directly to determine Hg° and dialkylmercury species. The other aliquot is derivatised with butylmagnesium chloride (a Grignard reagent) and chromatographed to determine Hg^{2+} and polar monoalkylated species.[15]

Because of the risk of artefacts it is recommended that a mercury-free gas condensate is analysed having a hydrocarbon composition as close as possible to that of the analysed sample as an analytical blank. The method of standard addition is a useful tool to confirm the identity of the peaks and to correct for signal suppression/enhancement errors. The recovery tests are essential since no certified reference materials exist for mercury species in natural gas and gas condensates. Calibration with solutions containing a single species is recommended to avoid errors stemming from species interconversion reactions.

Arsenic in Natural Gas and Gas Condensates

Arsenic species present in natural gas and shale oil are likely to be of biogeochemical origin. Indeed, the ability of marine organisms to bioaccumulate arsenic has been speculatively extended on paleoorganisms from which petroleum products were formed.[16] The organoarsenic compounds initially present in animals (*cf.* Chapter 13) may have undergone a series of alkylation-dealkylation and decarboxylation reactions to form volatile trialkyl arsines soluble in organic matrices.[16,17]

One individual organoarsenic species trimethylarsine accounts for more than 80% of total arsenic in some natural gases.[16] Organoarsines show a higher stability, even in air, in comparison with arsine (AsH_3). A GC chromatogram with atomic emission detection (GC-MIP AED) of a Me_3As standard blended with natural gas has been reported with a detection limit of *ca.* 5 pg.[2] Cryogenic sampling techniques are necessary to reduce hydrocarbon interferences. Low-temperature GC-MS (column maintained at 0 °C) has been used for the determination of trimethyl-, ethyldimethyl-, diethylmethyl-, and triethylmethylarsines; no chromatograms have, however, been reported.[16] The retention times are up to 1 h and the detection limits of 0.5 ng are insufficient for the analysis of real life samples.[16]

GC-ICP MS offers detection limits that are 100–10 000 lower in comparison with the above techniques. The direct injection of a gas condensate sample onto a capillary GC column coupled to ICP MS allows detection limits down to 0.05 pg to be obtained.[18] Arsenic is present as a number of species of which the two most abundant elute close to the retention time of triethyl arsine (Et_3As, b.p. 147 °C) and triphenylarsine (Ph_3As, b.p. 360 °C). The arsenic speciation is found to undergo a slight evolution as a function of sample storage time. Spiking with standards is found to change the equlibria in the system.[18]

A GC-ICP MS chromatogram shown in Figure 12.2 shows a stable baseline on the ^{75}As channel allowing the peaks of organoarsenic compounds to be distinguished. The ^{13}C signal can be used to monitor the elution of hydrocarbons whereas the Xe signal serves to monitor the plasma stability.[18] The problems concern the identification of the peaks in the chromatogram because of the unavailability of standards, and quantification because of the reactivity of spikes with other compounds present in the gas condensate matrix.

Metalloporphyrins in Coal and Shale Oil

Geoporphyrins usually occur as nickel (Ni^{2+}) and vanadyl (VO^{2+}) complexes in oils, oil shales, coals and sedimentary rocks in concentrations exceeding 10 $\mu g\,g^{-1}$. Other metals forming the coordination center include iron (Fe^{3+}), gallium (Ga^{3+}), manganese (Mn^{2+}) and titanyl (TiO^{2+}). Most studies of geoporphyrins have focused on structural elucidation of the demetallated porphyrin macrocycle.

Size exclusion chromatography with ICP MS detection has been proposed to study the molecular size distribution of specific elements in petroleum crudes.[19] For the separation of metallated porphyrins reversed phase HPLC[20,21] and high temperature capillary GC-ICP MS[21,22] have been proposed. Reversed phase HPLC using gradient elution with 0.2% pyridine in acetonitrile (1.0 to 1.5 ml min^{-1}) allows good resolution of nickel porphyrins.[20]

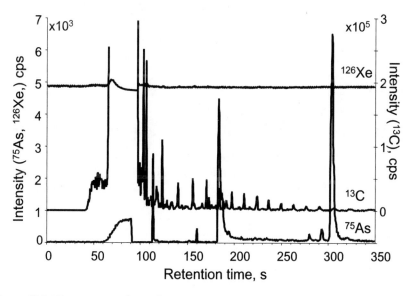

Figure 12.2 *Speciation analysis of arsenic in a gas condensate sample by GC-ICP MS* (from Ref. 18)

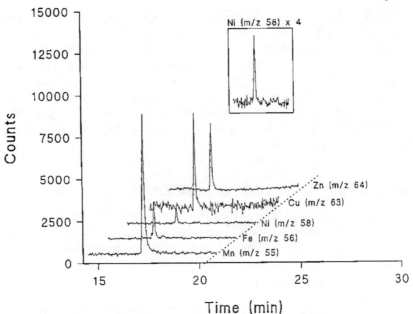

Figure 12.3 *A multielement chromatogram obtained by high temperature GC of metal-metalloporphyrin complexes*
(Reprinted from *J. Chromatogr. A*, 1993, **646**, 369, copyright 1993, with permission from Elsevier)

Atomic emission detection,[23-26] electron impact MS[27,28] and ICP MS[21,22] have been proposed for successful speciation analysis of metalloporphyrins by capillary GC. A multielement chromatogram obtained with high temperature intact metalloporpyrin complexes using a dedicated interface is shown in Figure 12.3.[22] The detection limits reported were at the 0.5 ng level but were obtained with early generation of ICP MS. A rapid degradation of the column efficiency was observed.[21]

5 Organolead and Organomanganese Species in Petrol

Organolead compounds are still added to petrol to improve its octane rating in some countries. The addition is usually a mixture of tetraalkyllead compounds. The levels of 0.15 mg L^{-1} are so high that a dilution of the sample, usually with hexane, is necessary. Capillary GC-AED and GC-ICP MS are equally good for the analysis. A chromatogram can be obtained within 5–10 min but multicapillary GC allows the baseline resolution of peaks to be obtained in less than 1 min.[29] A typical chromatogram is shown in Figure 12.4.

Methyl cyclopentadienylmanganese tricarbonyl has been used as a partial substitute for alkyllead compounds to improve the petrol octane rating. Although the compound is known to decompose in air, the presence of its residues in the atmosphere near petrol stations, in indoor car parks and at places where it is

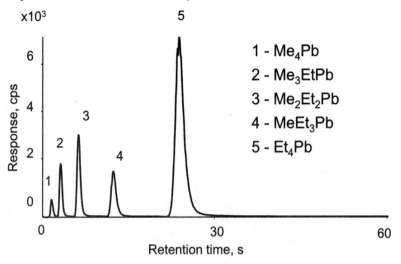

Figure 12.4 *A multicapillary GC-ICP MS chromatogram of organolead species in petrol* (from Ref. 29)

handled requires highly sensitive and specific methods for its regular monitoring.[30] GC-MIP AED has been reported to offer a detection limit of 0.5 fg which allows the detection of this compound and its analogues in environmental samples.[5,30]

References

1. V. Valcovic, *Trace Elements in Petroleum,* Petroleum Publishing Co., 1978.
2. S.S. Chao and A. Attari, *Characterization and measurement of natural gas trace constituents,* Contract N° 5089-253-1832, Gas Research Institute, Illinois, 1995.
3. P. Sarrazin, C.J. Cameron, Y. Barthel and M.E. Morrison, *Oil Gas J.,* 1993, **91**, 86 .
4. A. Veysseyre, K. van de Velde, C. Ferrari and C. Boutron, *Sci. Tot. Environ.,* 1998, **221**, 149.
5. H.B. Swan, Bull. Environ. Contam. Toxicol., 1999, **63**, 491.
6. C.J. Boreham and C.J.R. Fookes, *J. Chromatogr. A*, 1989, **467**, 195.
7. P. Sundararaman, *Anal. Chem.*, 1985, **57**, 2204.
8. S.M. Wilhelm and N. Bloom, *Fuel Proc. Technol.*, 2000, **63**, 1.
9. W. Frech, D.C. Baxter, B. Bakke, J. Snell and Y. Thomassen, *Anal. Commun.,* 1996, **33**, 7H.
10. N.S. Bloom, *Fresenius' J. Anal. Chem.*, 2000, **366**, 438.
11. C. Schickling and J.A.C. Broekaert, *Appl. Organomet. Chem.*, 1995, **9**, 29.
12. J. Snell, J. Qian, M. Johansson, K. Smit and W. Frech, *Analyst*, 1998, **123**, 905.

13. B. Bouyssiere, F. Baco, L. Savary and R. Lobinski, *Oil Gas Sci. Technol.*, 2000, **55**, 639.
14. H. Tao, T. Murakami, M. Tominaga and A. Miyazaki, *J. Anal. At. Spectrom.*, 1998, **13**, 1085.
15. B. Bouyssiere, F. Baco, L. Savary, and R. Lobinski, *J. Chromatogr. A*, 2002, **976**, 431.
16. K.J. Irgolic, D. Spall, B.K. Puri, D. Ilger and R.A. Zingaro, *Appl. Organomet. Chem.*, 1991, **5**, 117.
17. W. Delgado-Morales, M.S. Mohan and R.A. Zingaro, *Intern. J. Environ. Anal. Chem.*, 1994, **54**, 203.
18. B. Bouyssiere, F. Baco, L. Savary, H. Garaud, D.L. Gallup, and R. Lobinski, *J. Anal. At. Spectrom.*, 2001, **16**, 1329.
19. D.W. Hausler, *Spectrochim. Acta*, 1985, **40B**, 389.
20. K. Saitoh, H. Tanji, Y. Zheng, *Anal. Sci.*, 2001, **17**, i1511.
21. L. Ebdon, E.H. Evans, W.G. Pretorius and S.J. Rowland, *J. Anal. At. Spectrom.*, 1994, **9**, 939.
22. W.G. Pretorius, L. Ebdon and S.J. Rowland, *J. Chromatogr. A*, 1993, **646**, 369.
23. B.D. Quimby, P.C. Dryden and J.J. Sullivan, *J. High Res. Chromatogr.*, 1991, **14**, 110.
24. D.W. Hausler and D.H. Renfro, *Prepr. Am. Chem. Soc., Div. Petr. Chem.*, 1991, **36**, 225.
25. Y. Zeng and PC. Uden, *J. High Res. Chromatogr.*, 1994, **17**, 223.
26. Y. Zeng and P.C. Uden, *J. High Res. Chromatogr.*, 1994, **17**, 217.
27. W. Blum, P. Ramstein and G. Eglinton, *J. High Res. Chromatogr.*, 1990, **13**, 85.
28. W. Blum, W.J. Richter and G. Eglinton, *J. High Res. Chromatogr.*, 1988, **11**, 148.
29. I. Rodriguez Pereiro and R. Lobinski, *J. Anal. At. Spectrom.*, 1997, **12**, 1381.
30. Y.K. Chau, F. Yang and M. Brown, *Appl. Organomet. Chem.*, 1997, **11**, 31.

CHAPTER 13

Speciation of Redox States

1 Introduction

Several elements, such as arsenic, antimony, chromium, selenium or tellurium, occur at different, usually two, principal oxidation states, one of which is toxic and the other not. The differentiation between these oxidation states is one of the principal areas of speciation analysis. Many analytical methods for the discrimination of redox states have a long history and are not chromatographic. Elements, such as As, Se or Sb are known to form hydrides only at the lower oxidation states. Therefore hydride generation AAS or ICP MS carried out directly on the sample allows the specific determination of As(III), Se(IV) or Sb(III) in the presence of As(V), Se(VI) or Sb(V), respectively. The elements at the higher oxidation states are determined by the difference between the total element concentration (found using the same technique after reduction) and the concentration of the lower oxidation state. Another group of techniques is based on flow injection analysis. A sample containing an element at two oxidation states is injected onto a solid phase extraction membrane or a chromatographic microcolumn, able to retain one of the species of interest. The non-retained species is determined directly, the retained one is subsequently eluted using different chemical conditions.

A discussion of the non-chromatographic approaches for speciation analysis of redox states is beyond the scope of this book and is reviewed elsewhere.[1] This chapter focuses on methods based on the chromatographic separation of the different redox states, either native or complexed to form chelates separable by liquid chromatography.

2 Analytical Methodology

Redox species can be separated in native states or one or both can be derivatised in order to produce chromatographable species. Another important consideration is the tendency of some elements to hydrolyse which requires the presence of a complexing agent in the mobile phase. The problems to be solved during method development include the need to maintain the species stability over the timescale of the chromatographic or electrophoretic run, and control polyatomic interferences in ICP MS, especially in the analysis of carbon or chlorine-rich matrices, such as humic-rich samples or seawater, respectively.

Anion exchange HPLC-ICP MS is a versatile analytical technique for speciation analysis of redox states. The analysis time can be reduced by the use of microcolumns.[2] An emerging alternative is capillary zone electrophoresis.[3] Efficient multielement separations of metal states in environmental matrices followed by on-line ICP MS detection have been reported.[4-7] Examples of multispecies separations carried out by HPLC-ICP MS[6] and CZE-ICP MS[4] are shown in Figure 13.1.

The significance of reversed phase HPLC-ICP MS is declining since a

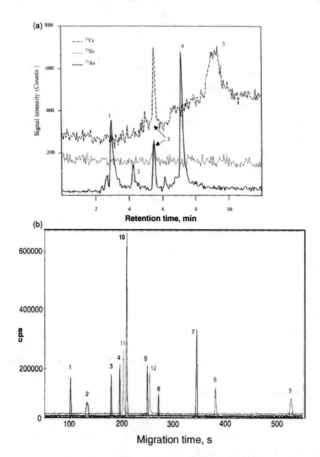

Figure 13.1 *Examples of multielemental multispecies separations by: a) anion exchange HPLC-ICP MS [see text for peak identification] (from Ref. 6); and b) CZE-ICP MS [1 – arsenocholine, 2 – arsenobetaine, 3 – arsenite, 4 – dimethylarsonic acid, 5 – phenylarsonic acid, 6 – monomethylsarsinic acid, 7 – arsenate, 8 – selenite, 9 – selenate, 10 – antimonate, 11 – tellurite, 12 – tellurate]* (courtesy of Dr. Andreas Prange)

derivatisation reaction for at least one redox state, and hence a change in the equilibrium, is required.

3 Overview of Applications

Antimony

The inorganic species of Sb are more toxic than the methylated ones, and Sb(III) is 10 times more toxic than Sb(V). Sb(V) is predominant in oxygenated waters while Sb(III) dominates in reducing interstitial waters. Both oxidation states are readily hydrolysed. Analytical methods for the determination of antimony species have been reviewed.[8]

In aqueous solution, Sb(V) is present as a monocharged anion in the pH range 2.7 to 10.4. Under these conditions Sb(III) hydrolyses and precipitates on the column. A chromatographic separation of Sb(V) from Sb(III) requires the stabilisation of the latter in solution. It can be achieved with EDTA leading to the Sb(EDTA)$^-$ anion. Sb(III) and Sb(V) can be separated by anion exchange chromatography using EDTA containing phthalate as the mobile phase.[9,10] Because of the relatively low Sb concentrations in samples of interest ICP MS is the only detection technique capable of producing valid results. Detection limits can be decreased by using hydride generation.[11] Still lower detection limits, down to the 10 pg mL^{-1} have been reported using an ultrasonic nebuliser.[12]

Sb(III) is easily oxidised to Sb(V) within few hours. Contamination by leaching from glass bottles, especially under alkaline conditions, has been reported.[12] The use of polyethylene bottles and plastic syringes for injection is essential.

Arsenic and Selenium

The redox states of arsenic and selenium have been the species most frequently determined by hyphenated techniques. Arsenic, highly toxic as As(III), can be differentiated from As(V) and organoarsenic compounds which show far less toxicity. Inorganic arsenic is predominant in ground water whereas monomethylarsinic acid and dimethylarsonic acid are the most commonly detected species in marine waters. The essential Se(IV) needs to be differentiated from Se(VI); trimethylselenonium cation, selenomethionine and selenocysteine are often included in separations, especially in extracts of biological materials.

Anion exchange chromatography run in acidic[13] or, more often alkaline media[2,14,15] allows the separation of six As species (methylarsinic acid, arsenous acid, dimethylarsonic acid, arsenic acid, arsenobetaine and arsenocholine) and five Se species (selenite, trimethylselenonium cation, selenocystine, selenomethionine and selenate) within one run. ICP MS is the detection technique of choice. Typical detection limits are 0.1 and 0.5 μg L^{-1} for As and Se species, respectively. ^{82}Se isotope should be monitored in order to avoid chloride interference, which may occur in the case of seawater analysis. In the case of arsenic, chloride is separated from arsenic species and can be observed in the chromatogram as a peak. The use of shield torch and collision cell instruments considerably reduces any polyatomic interference.[16]

Bromine and Iodine

The trace analysis of bromate is becoming increasingly important as this ion is believed to be carcinogenic and may be formed during oxidative processes in drinking water. The technique of choice is ion chromatography-ICP MS which allows a rapid and sensitive determination of all bromine species with a DL of $60 \, \text{ng} \, \text{L}^{-1}$ within 10 min.[17–19] The limit of detection is restricted by contamination of the plasma argon with bromine. $^{40}\text{Ar}_2\text{H}^+$ molecular ion strongly interferes with the ^{81}Br, resulting in impaired accuracy when using quadrupole mass spectrometry. An HPLC isotope dilution ICP MS method for the accurate determination of bromate has been developed.[19]

The inorganic iodine species iodide and iodate can also be separated by ion chromatography prior to ICP MS detection.[20] On-line isotope-dilution ICP MS has been used for species quantification.[20]

Chromium

The different bioavailability and toxicity of chromium species (Cr(VI) is more toxic than Cr(III)) has occupational health implications. The analytical techniques of choice include anion exchange and reversed phase HPLC with ICP MS detection. Chromium speciation in liquid samples has been reviewed.[21]

Cr(III) and Cr(VI) can be separated by anion exchange HPLC using 0.25% HNO_3 as the mobile phase.[22] Chromium(III) can be stabilised with EDTA to form an anion separated from Cr(VI).[23] For reversed phase HPLC, complexes formed by Cr(III) and Cr(VI) with pyrrolidine-1-carbodithioate have been preconcentrated by SPE on C_{18}-bonded silica and separated by reversed phase HPLC with aqueous acetonitrile as the mobile phase.[24] Alternatively ion interaction reversed phase HPLC (using a tetrabutylammonium cation in the mobile phase) has been proposed.[25] Detection limits of $1 \, \text{ng} \, \text{mL}^{-1}$ can be readily obtained.

The problems of species interconversion are pertinent. The analysis should be performed as soon as possible after sampling to avoid species degradation and interconversion. An extensive study recommends storage in quartz vessels at $5 \, ^\circ\text{C}$ at pH 6.4 with HCO_3/H_2CO_3 under a CO_2 blanket to avoid Cr(VI) reduction, owing to its high oxidising strength at low pH values.[26]

Accurate quantification in speciation analysis of redox states of Cr with a quadrupole mass spectrometer is hampered by polyatomic interferences from carbon- ($^{40}\text{Ar}^{12,13}\text{C}^+$) and/or chlorine-based ($^{37}\text{Cl}^{16}\text{O}^+$ and $^{35}\text{Cl}^{16}\text{OH}^+$) molecular ions. The effect of these interferences on the accuracy has been exhaustively studied.[27] The use of a medium resolution MS (R = 3000) is recommended over cool plasma conditions for the efficient removal of the interference.[27] Speciated isotope dilution MS has been proposed to improve the accuracy of Cr speciation.[28]

The interest in routine chromium speciation has spured the development of automated analysers. Sample introduction systems based on automated LC[29] and on microchip capillary electrophoresis[30] have been proposed to enable speciation analysis for chromium by ICP MS.

References

1. A.K. Das, M. de la Guardia and M.L. Cervera, *Talanta*, 2001, **55**, 1.
2. A. Woller, H. Garraud, J. Boisson, A.M. Dorthe, P. Fodor and O.F.X. Donard, *J. Anal. At. Spectrom.*, 1998, **13**, 141.
3. E. Dabek Zlotorzynska, E.P.C. Lai and A.R. Timerbaev, *Anal. Chim. Acta*, 1998, **359**, 1.
4. A. Prange and D. Schaumlöffel, *J. Anal. At. Spectrom.*, 1999, **14**, 1329.
5. D. Schaumlöffel and A. Prange, *Fresenius' J. Anal. Chem.*, 2000, **364**, 452.
6. Y. Martinez-Bravo, A.F. Roig-Navarro, F.J. Lopez and F. Hernandez, *J. Chromatogr. A*, 2001, **926**, 265.
7. C. Casiot, O.F.X. Donard and M. Potin-Gautier, *Spectrochim. Acta*, 2002, B**57**, 173.
8. P. Smichowski, Y. Madrid and C. Camara, *Fresenius' J. Anal. Chem.*, 1998, **360**, 623.
9. J. Lintschinger, I. Koch, S. Serves, J. Feldmann and W.R. Cullen, *Fresenius' J. Anal. Chem.*, 1997, **359**, 484.
10. J. Lintschinger, O. Schramel and A. Kettrup, *Fresenius' J. Anal. Chem.*, 1998, **361**, 96.
11. P. Smichowski, Y. Madrid, M.B. de la Calle Guntinas and C. Camara, *J. Anal. At. Spectrom.*, 1995, **10**, 815.
12. M. Krachler and H. Emons, *Anal. Chim. Acta*, 2001, **429**, 125.
13. J. Zheng and W. Kosmus, *J. Liq. Chromatogr. Relat. Technol.*, 1998, **21**, 2831.
14. T. Guerin, A. Astruc, M. Astruc, A. Batel and M. Borsier, *J. Chromatogr. Sci.*, 1997, **35**, 213.
15. B.P. Jackson and W.P. Miller, *Environ. Sci. Technol.*, 1999, **33**, 270.
16. T. Nakazato, H. Tao, T. Tanigushi and K. Isshiki, *Talanta*, 2002, **58**, 121.
17. M. Nowak and A. Seubert, *Anal. Chim. Acta*, 1998, **359**, 193.
18. B. Divjak, M. Novic and W. Goessler, *J. Chromatogr. A*, 1999, **862**, 39.
19. A. Seubert and J. Godfrey, *Int. Lab. News*, 2001, 18.
20. K.G. Heumann, S.M. Gallus, G. Rädlinger and J. Vogl, *Spectrochim. Acta*, 1998, **53B**, 273.
21. M.J. Marqués, A. Salvador, A. Morales-Rubio and M. de la Guardia, *Fresenius' J. Anal. Chem.*, 2000, **367**, 601 .
22. M.J. Powell, D.W. Boomer and D.R. Wiederin, *Anal. Chem.*, 1995, **67**, 2474.
23. F.A. Byrdy, L.K. Olson, N.P. Vela and J.A. Caruso, *J. Chromatogr. A*, 1995, **712**, 311.
24. C.M. Andrle, N. Jakubowski and J.A.C. Broekaert, *Spectrochim. Acta*, 1997, **52B**, 189.

25. N. Jakubowski, B. Jepkens, D. Stuewer and H. Berndt, *J. Anal. At. Spectrom.*, 1994, **9**, 193.
26. S. Dyg, R. Cornelis, B. Griepink and P. Quevauviller, *Anal. Chim. Acta*, 1994, **286**, 297.
27. F. Vanhaecke, S. Saverwyns, G. De Wannemacker, L. Moens and R. Dams, *Anal. Chim. Acta*, 2000, **419**, 55.
28. H.M. Kingston, D. Huo, Y. Lu and S. Chalk, *Spectrochim. Acta*, 1998, **53B**, 299.
29. M.K. Donais, R. Henry and T. Rettberg, *Talanta*, 1999, **49**, 1045.
30. Q.J. Song, G.M. Greenway and T. McCreedy, *J. Anal. At. Spectrom.*, 2003, **18**, 1.

CHAPTER 14

Speciation of Organoarsenic Compounds in Biological Materials

1 Introduction

Arsenic has the notoriety of being a toxic element but its toxicity is critically dependent on the chemical form in which it occurs. Speciation of arsenic in food is of particular interest due to the potential accumulation of arsenic in the food chain and the risk to man. The metabolism of inorganic arsenic by marine and terrestrial plants and animals leads to the formation of a range of organic arsenic species that may be considered as naturally occurring compounds.[1] The most widely studied of this group is a quaternary arsenocompound, arsenobetaine, which is the major organoarsenic compound in marine animals, and arsinoylribo-sides (arsenosugars) which are products of As metabolism in marine plants and some bivalves. Another field of interest is the study of the metabolism of arsenic following its administration to humans and experimental animals by means of the determination of arsenic speciation in urine. The most important arsenic species of interest in speciation analysis are summarised in Figure 14.1. Besides compounds with a covalent As-C bond, As(III) may form complexes with thiol groups of glutathione (GSH) or proteins.[2]

The literature on the speciation of arsenic is extremely rich with more than 1000 papers available. In most of these, the commercially available standards, As(III), As(V), monomethylarsinic (MMAA) and dimethylarsonic acids (DMAA), and later arsenobetaine and arsenocholine have been used with the aim of optimising a method, often without the objective of applying it to a real life sample. An overview of these techniques can be found in several extensive reviews.[3-5] Other studies of arsenic speciation in marine life are based on multistep chromatographic purification prior to NMR identification developed by Francesconi and Edmonds.[6] The downscaling and refining of these methods have

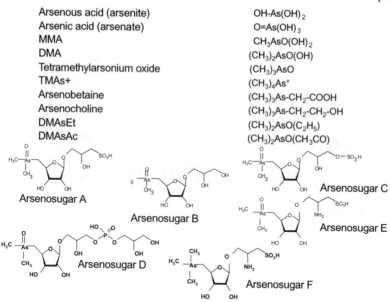

Arsenous acid (arsenite)	$OH\text{-}As(OH)_2$
Arsenic acid (arsenate)	$O=As(OH)_3$
MMA	$CH_3AsO(OH)_2$
DMA	$(CH_3)_2AsO(OH)$
Tetramethylarsonium oxide	$(CH_3)_3AsO$
TMAs+	$(CH_3)_4As^+$
Arsenobetaine	$(CH_3)_3As\text{-}CH_2\text{-}COOH$
Arsenocholine	$(CH_3)_3As\text{-}CH_2\text{-}CH_2\text{-}OH$
DMAsEt	$(CH_3)_2AsO(C_2H_5)$
DMAsAc	$(CH_3)_2AsO(CH_3CO)$

Figure 14.1 *Arsenic species in biological materials*

resulted in several analytical procedures employing the coupling of HPLC, usually with ICP MS.

HPLC-ICP MS is a well established technique for routine monitoring of arsenic species in biological fluids and biological samples extracts. Alternative techniques, either lack the sensitivity (*e.g.* ICP AES) or include a post-column chemical reaction step (HPLC-AAS or HPLC-AFS) that results in a complexity and unsuitability for routine applications.

Because of the poor availability of standards, electrospray MS/MS is becoming an essential complementary analytical technique for arsenic speciation studies. Arsenic has only one stable isotope. Therefore, the recognition of a peak belonging to an arsenocompound in a mass spectrum requires either the accurate measurement of the molecular mass (using a TOF or, better, an FT ICR mass analyser), or the fragmentation of all the peaks in the mass spectrum looking for characteristic fragments in product ion mass spectra.

This chapter overviews the status of HPLC-ICP MS for the determination of arsenic species in biological materials, purification protocols prior to electrospray MS and the standardless identification and characterisation of unknown organoarsenic species by ES MS/MS.

2 Determination of Arsenic Species by HPLC-ICP MS

Separation of Arsenic Species

HPLC separation mechanisms take advantage of the fact that organoarsenic compounds are readily and differently ionised at different pH owing to the presence of functional groups such as $Me_2As(O)$- or $-SO_3H$. The ions formed (cations from the protonation of the dimethylarsinoyl moiety at low pH and anions from the dissociation of the $-SO_3H$ groups at neutral and basic pH) are separated by ion exchange HPLC. Alternatively, they can be made to form hydrophobic compounds with a suitable ion-pairing reagent and are then chromatographed by RP HPLC.

Anion exchange HPLC-ICP MS has the best potential for comprehensive speciation analysis of organoarsenic species in biological materials. Isocratic elution in neutral media has been recommended by Larsen for the separation of the five major arsenic species (As(III), As(V), MMAA, DMAA, and arsenobetaine).[7] The isocratic elution mode and the low concentrations of organic solvent and dissolved solids in the mobile phases permit stable operation of the ICP with no degradation in sensitivity and little matrix interference. A set of 14 guidelines for performing speciation of arsenic by HPLC-ICP MS has been proposed.[7] A 1.5 cm C_{18} guard column placed before the AE column has been used to remove most of the organic components from urine that would otherwise bind irreversibly to the packing material.[8] An example of a comprehensive speciation of arsenic species by anion exchange HPLC is shown in Figure 14.2.[9]

Cation exchange HPLC has often been used in parallel with AE HPLC to confirm the identity of the analyte species, especially for samples that might contain arsinoylribosides. Most of organoarsenic species are not retained on a reversed phase column; the addition of an ion pairing reagent is required to achieve their separation. Both anion pairing cations such as tetraethyl ammonium[10–12] or tetrabutylammonium,[13,14] and cation pairing anions such as pentanesulfonate,[15] hexanesulfonate,[10] heptanesulfonate[16,17] and dodecyl sulfonate[13,18] have been used for this purpose.

Another separation mechanism frequently applied is based on size exclusion chromatography of ion pairs of organoarsenic compounds including arsenosugars.[11,12,19,20] The use of a size exclusion column with 1% acetic acid as the mobile phase has been shown to be capable of the baseline separation of 12

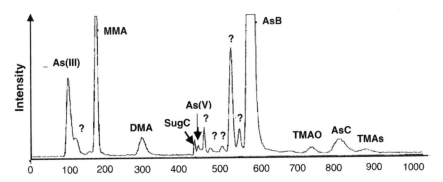

Figure 14.2 *Anion exchange HPLC-ICP MS chromatogram of organoarsenic species in an oyster tissue*
(reprinted with permission from *Rapid Commun. Mass Spectrom.*, 2002, **16**, 965, copyright 2002, John Wiley & Sons)

organoarsenic compounds into four fractions.[21] Size exclusion chromatography is a convenient clean-up technique for organoarsenic species prior to their final separation by reversed phase or anion exchange HPLC. Size exclusion LC has been recommended for the isolation of the seleno-bis(*S*-glutathionyl)arsinium ion in rabbit bile.[22]

Detection of Arsenic Species

ICP MS coupled *via* a conventional cross-flow or concentric nebuliser is the established detection technique in speciation analysis of arsenic in biological materials. The signal-to-noise ratio can be improved four-fold by the introduction of 3% (*v/v*) of methanol into the mobile phase and increasing the ICP power.[23] Lower detection limits $(0.04-0.6 \ \mu g \, L^{-1})$ have been obtained by the use of a hydraulic high pressure nebuliser[24,25] or by the conversion of the organoarsenic compounds exiting the column into AsH_3.[26,27] The latter procedure also alleviates interference from the polyatomic ion $^{40}Ar^{35}Cl^+$ at *m/z* 75 which may occur during the analysis of samples with a high chloride content, such as urine.

Interest in routine speciation analysis has resulted in a number of interfaces allowing the detection of organoarsenic compounds by AAS and AFS. The prerequisite for success is the conversion of organoarsenic species into AsH_3. This can be achieved thermochemically,[28] by microwave-assisted oxidation[29-31] or by UV-photooxidation.[32,33] The detection limits then match those achieved in HPLC-ICP MS with pneumatic nebulisation. They are in the low $\mu g \, L^{-1}$ range but are species dependent.

Sample Preparation of Marine Biota and Seafood Samples

Defatting (removal of lipids), *e.g.* by leaching with acetone,[34] diethyl- or petroleum ether[35] is usually the first analytical step in the speciation of arsenic in animal tissues and marine foodstuffs. A mass balance should be verified after extraction since some arsenic compounds are fat soluble and can be lost during defatting. It is also necessary to avoid generating an emulsion with the fat, which would make the subsequent clean-up more difficult.[26] Since some samples of seafood products are prepared in oil and generally tend to have a high salt content an additional clean-up step, *e.g.* on a strong cation exchanger[34] is required to eliminate the remains of liposoluble compounds not removed during the defatting step. The clean-up also eliminates pressure buildup due to accumulation of material on the column. In addition, the efficiency of the subsequent methanol extraction step is apparently higher for defatted samples than for non-defatted ones.[34] Uncleaned samples may also generate problems with the ICP MS cones and ES ion source.

Quantitative separation of organoarsenic from the matrix is desirable for a chromatographic analysis. Extraction of arsenobetaine, arsenocholine and arsinoylribosides is usually performed using methanol,[34] methanol-chloroform-water[36,37] or methanol-water.[38,39] A comparison study of these methods is available.[40] A methanol-water mixture is recommended for the dry tissues whereas fresh samples can be efficiently leached with pure methanol. Recoveries reach 90% for fish and 80% for mussels. No degradation of arsenobetaine to other species was observed when an enzymic (trypsin) digestion procedure was applied to the fish.[31,41] The methanolic extraction is typically repeated 2–3 times followed by preconcentration of the extract by evaporation of methanol using a rotavaporator. Arsenobetaine in sample extracts that were stored at 4 °C for 9 months was reported to decompose to trimethylarsine oxide and two other unidentified arsenic species.[16,42]

There has been a surge of interest recently in the use of microwave-assisted procedures for the recovery of organoarsenic compounds from biological tissues.[15,43,44] The use of an automated solvent accelerated extractor for the same purpose is found to be efficient.[45] The major advantage of these procedures is that the increased speed of leaching reduces the sample preparation time to a few minutes. It should be noted that application of the extraction methods developed for marine samples to samples of plants or terrestrial organisms is plagued by non-quantitative recoveries.

Sample Preparation of Urine Samples

Despite its apparent simplicity urine is a complex matrix but its constituents are water-soluble which allows direct chromatography. The sample may be troublesome to handle because of its higher salt concentration, in comparison with the concentrations of metallocompounds. Urine contains *ca.* 1% NaCl although the concentration varies greatly. The urine matrix causes column overloading which results in peak broadening.[46] High dilution factors may be necessary.[47,46]

Urine samples can be collected in polycarbonate bottles and filtered through a 0.45 μm filter.[8] The storage procedure implies the use of acid-washed PE bottles at −10 °C[46] but storage at 20 °C has also been reported.[8] Urine samples (filtered) can be injected on a chromatographic column directly or after dilution with diluted acid.[32,48–50] A freezing procedure has been commonly used to preserve biosamples; arsenic speciation in fresh and defrosted samples has been compared.[16]

3 Identification of Arsenic Species by Electrospray MS/MS

In the overwhelming majority of work the identification of signals in HPLC with As-specific detection has been achieved by matching their retention times with the standards. Retention time irreproducibility is common in the presence of a sample matrix of which some constituents may form ion pairs with organoarsenic compounds. Also, the condition of the column plays a role, the efficiency of separation decreasing slowly with the number of chromatographic runs on a given column. Some of these problems can be solved by spiking experiments but there is a risk that the spike of one compound will match the retention time of another one leading to signal misidentification. This risk can be reduced by the use of 2- or 3-dimensional chromatography.[21,51]

The availability of standards for organoarsenic compounds is still problematic although the situation has improved with the availability of pure arsenobetaine and arsenocholine calibrants from the Measurement and Testing Programme[52] and the National Institute of Environmental Studies (NIES).[11] Regarding arsenosugars, the only source remains their isolation from natural samples by preparative chromatography[6] or the use of well characterised reference materials.[11] The poor availability of standards has spurred the interest in electrospray MS/MS for speciation analysis of arsenic.

Multidimensional LC Purification of Arsenic Species Prior to ES MS/MS

Electrospray ionisation suffers from signal suppression by the matrix and a generally poor compatibility of the chromatographic mobile phase with the ionisation conditions. In particular, phosphate buffers used in anion exchange chromatography and ion pairing reagents used for the most efficient separations of organoarsenic compounds seem to be disastrous for the sensitivity of ES MS. Figure 14.3 demonstrates the need for multidimensional chromatographic purification of organoarsenic species prior to ES MS analysis, using the example of the identification of an arsinoylriboside (arsenosugar B in Figure 14.1).

A purification solely by size exclusion chromatography is insufficient to assure a detectable signal at m/z 329 where the arsenosugar is expected. The combination of SEC with anion exchange allows the identification of a small peak at m/z 329 which gives a correct collision induced dissociation fragmentation pattern (not shown). However, in order to attribute this particular peak to the arsenic compound of interest, a prior knowledge of its molecular mass is necessary. Only after a third dimension separation, by cation exchange HPLC, is the purity of the arsenosugar sufficient enough to assure its intensity to be dominant among other peaks in the mass spectrum.

Molecular Mass Determination of Arsenic Species by ES MS

ES MS or HPLC-ES MS can be used to confirm conclusions drawn on the basis of HPLC-ICP MS data[53,54] and also to identify novel compounds.[55-57] Detection limits down to 10 ng mL^{-1} can be achieved in capillary HPLC-ES MS. The basic problem in identification by ES MS is the difficulty in recognising the arsenic peaks in a mass spectrum in view of its being monoisotopic.

The unit resolution of a quadrupole filter is insufficient for the unambiguous confirmation of the compound identity on the basis of the molecular mass. A better resolution and higher mass measurement accuracy can be obtained by using a TOF mass analyser. The combination of time delayed extraction of ions with the use of an ion reflectron allows resolution above 10 000 to be obtained with a mass accuracy below 10 ppm. Figure 14.4 shows the effect of the resolution of

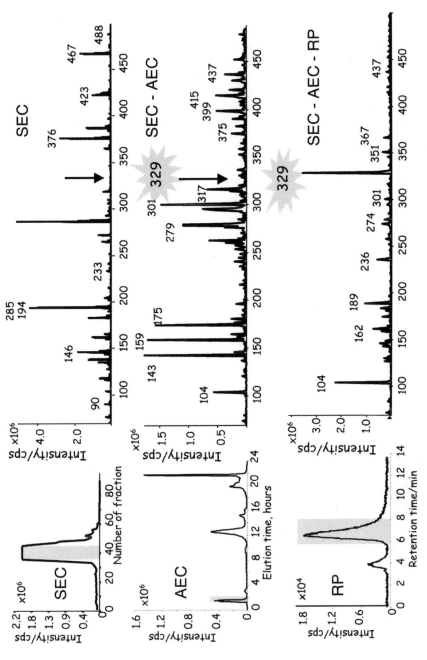

Figure 14.3 *Effect of the chromatographic purification of organoarsenic species on the quality of mass spectra in electrospray MS/MS analysis*

the mass analyser used for the identification of an arsenic peak in a mass spectrum. The molecular mass determined by TOF MS (Figure 14.4a) matches that calculated on the basis of an empiric formula $C_{10}H_{21}O_7As$ with an error of 35 ppm. The observed isotopic pattern of the molecular ion, taking into account the isotopic composition of all the elements present in the sample, closely matches that calculated on the basis of its molecular formula.

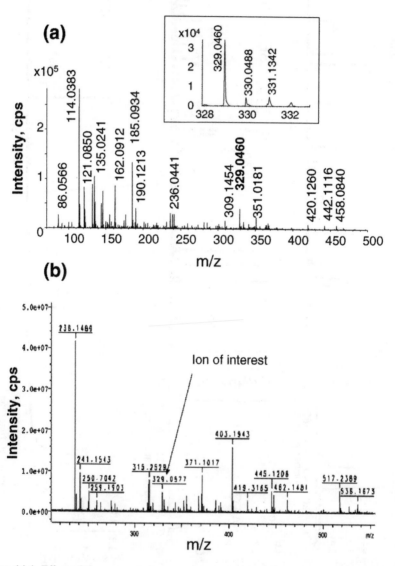

Figure 14.4 *Effect of the mass analyser on the accuracy of the molecular mass measurement: a) time-of-flight MS (zoom in the inset)* (from Ref. 57); *b) Fourier Transform ICR MS*
(from Ref. 58)

The error of the molecular mass measurement can be further minimised by improving the resolution and mass measurement accuracy which is possible using an FT ICR mass spectrometer (Figure 14.4b). In this way resolution exceeding 200 000 and mass accuracy down to 0.2 ppm can be obtained.[58]

The determination of the molecular mass, however exactly it is determined, is insufficient for structure elucidation, for which the use of tandem mass spectrometry is required.

Structural Characterisation of Arsenic Compounds by Multiple MS

The identity confirmation or identification of an unknown compound can be achieved by collision induced dissociation MS. This is a valuable tool for the speciation analysis of arsenic and facilitates a positive identification of the species detected. CID data can be acquired using triple quad (QqQ), quadrupole time-of-flight (QqTOF)[57–59] or ion trap[60,61] mass analysers. Figure 14.5 shows a tandem mass spectrum allowing the recognition of characteristic fragments of an arsinoylriboside and thus the attribution of the *m/z* 329 peak to a particular arsenosugar (arsenosugar B).[57] CID FT ICR mass spectra of some organoarsenic compounds have been reported.[58]

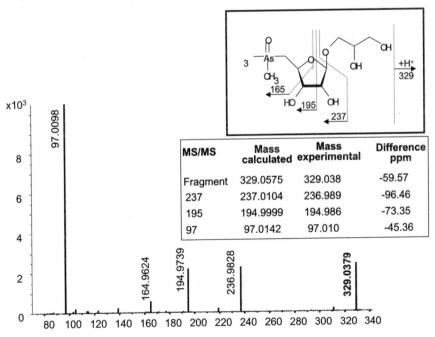

MS/MS	Mass calculated	Mass experimental	Difference ppm
Fragment	329.0575	329.038	-59.57
237	237.0104	236.989	-96.46
195	194.9999	194.986	-73.35
97	97.0142	97.010	-45.36

Figure 14.5 *Confirmation of the m/z 329 peak as belonging to arsenosugar B by electrospray Q-TOF MS*
(adapted from *J. Chromatogr. A*, 2001, **926**, 171, copyright 2001, with permission of Elsevier)

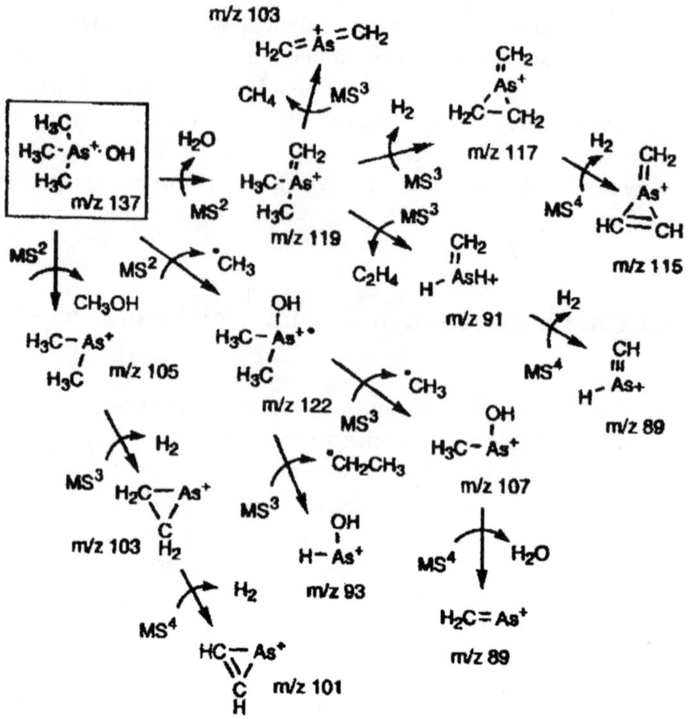

Figure 14.6 *Elucidation of the structure of an organoarsenic compound by ion trap MS[n]* (reprinted with permission from *J. Chromatogr. A*, 2001, **926**, 167, copyright 2001, Elsevier)

The full understanding of the fragmentation pathways of organoarsenic compounds requires multiple mass spectrometry. Multiple mass spectra (MS[n], n = 1–6) can be obtained using ion trap mass spectrometry. Figure 14.6 summarises information on the fragmentation pathways of an organoarsenic compound obtained using this technique.[61]

4 Validation of Arsenic Speciation Analysis

The fact that arsenic is monoisotopic renders isotope dilution analysis by ICP MS impossible. The analysis of a CRM is a unique way of method validation. BCR 627 Tuna fish and DORM-2 (NRC) Dogfish muscle are available with certified arsenobetaine concentrations of $52 \pm 3 \, \text{mmol} \, \text{kg}^{-1}$ and $16.4 \pm 1.1 \, \text{mg} \, \text{kg}^{-1}$ (as As), respectively. Other certified species include DMA at $2 \pm 0.3 \, \text{mmol} \, \text{kg}^{-1}$ in BCR 627 and TMAs$^+$ at $0.248 \pm 0.054 \, \text{mg} \, \text{kg}^{-1}$ as As in DORM-2.

References

1. M. Morita and J.S. Edmonds, *Pure Appl. Chem.*, 1992, **64**, 575.
2. K.T. Suzuki, B.K. Mandal and Y. Ogra, *Talanta*, 2002, **58**, 111.
3. M. Burguera and J.L. Burguera, *Talanta*, 1997, **44**, 1581.
4. T. Guerin, A. Astruc and M. Astruc, *Talanta*, 1999, **50**, 1.
5. Z.L. Gong, X.F. Lu, M.S. Ma, C. Watt and X.C. Le, *Talanta*, 2002, **58**, 77.
6. K. Francesconi and J. Edmonds, *Oceanogr. Mar. Biol. Annu. Rev.*, 1993, **31**, 111.
7. E.H. Larsen, *Spectrochim. Acta*, 1998, **53B**, 253.
8. B.S. Chana and N.J. Smith, *Anal. Chim. Acta*, 1987, **197**, 177.
9. U. Kohlmeyer, J. Kuballa and E. Jantzen, *Rapid Commun. Mass Spectrom.*, 2002, **16**, 965.
10. X.C. Le, X.F. Li, V. Lai, M. Ma, S. Yalcin and J. Feldmann, *Spectrochim. Acta*, 1998, **53B**, 899.
11. J. Yoshinaga, Y. Shibata, T. Horiguchi and M. Morita, *Accred. Qual. Assur.*, 1997, **2**, 154.
12. Y. Shibata and M. Morita, *Anal. Sci.*, 1989, **5**, 107.
13. D. Beauchemin, K.W.M. Siu, J.W. McLaren and S.S. Berman, *J. Anal. At. Spectrom.*, 1989, **4**, 285.
14. M.B. Amran, F. Lagarde and M.J.F. Leroy, *Mikrochim. Acta.*, 1997, **127**, 195.
15. K.L. Ackley, C. B'Hymer, K.L. Sutton and J.A. Caruso, *J. Anal. At. Spectrom.*, 1999, **14**, 845.
16. S.X.C. Le, W.R. Cullen and K.J. Reimer, *Environ. Sci. Technol.*, 1994, **28**, 1598.
17. K.A. Francesconi, P. Micks, R.A. Stockton and K.J. Irgolic, *Chemosphere*, 1985, **14**, 1443.
18. D. Beauchemin, M.E. Bednas, S.S. Berman, J.W. McLaren, K.W.M. Siu and R.E. Sturgeon, *Anal. Chem.*, 1988, **60**, 2209.
19. Y. Shibata and M. Morita, *Anal. Chem.*, 1989, **61**, 2116.
20. M. Morita and Y. Shibata, *Anal. Sci.*, 1987, **3**, 575.
21. S. McSheehy and J. Szpunar, *J. Anal. At. Spectrom.*, 2000, **15**, 79.
22. J. Gailer, S. Madden, G.A. Buttigieg, M.B. Denton and H.S. Younis, *Appl. Organomet. Chem.*, 2002, **16**, 72.
23. E.H. Larsen and S. Sturup, *J. Anal. At. Spectrom.*, 1994, **9**, 1099.
24. J. Zheng, W. Goessler and W. Kosmus, *Mikrochim. Acta.*, 1998, **130**, 71.
25. W. Goessler, W. Maher, K.J. Irgolic, D. Kuehnelt, C. Schlagenhaufen and T. Kaise, *Fresenius' J. Anal. Chem.*, 1997, **359**, 434.
26. T. Dagnac, A. Padro, R. Rubio and G. Rauret, *Talanta.*, 1999, **48**, 763.
27. D. Beauchemin, *J. Anal. At. Spectrom.*, 1998, **13**, 1.
28. J.S. Blais, G.M. Momplaisir and W.D. Marshall, *Anal. Chem.*, 1990, **62**, 1161.
29. D. Velez, N. Ybanez and R. Montoro, *J. Anal. At. Spectrom.*, 1997, **12**, 91.
30. D.L. Tsalev, M. Sperling and B. Welz, *Analyst*, 1992, **117**, 1735.
31. K.J. Lamble and S.J. Hill, *Anal. Chim. Acta*, 1996, **334**, 261.
32. D.L. Tsalev, M. Sperling and B. Welz, *Analyst*, 1998, **123**, 1703.

33. X. Zhang, R. Cornelis, J. De Kimpe and L. Mees, *Anal. Chim. Acta.*, 1996, **319**, 177.
34. N. Ybanez, D. Velez, W. Tejedor and R. Montoro, *J. Anal. At. Spectrom.*, 1995, **10**, 459.
35. K. Shiomi, Y. Sugiyama, K. Shimakura and Y. Nagashima, *Appl. Organomet. Chem.*, 1995, **9**, 105 .
36. E. H. Larsen, G. Pritzl and S.H. Hansen, *J. Anal. At. Spectrom.*, 1993, **8**, 1075.
37. E.H. Larsen, *Fresenius' J. Anal. Chem.*, 1995, **352**, 582.
38. V.W.-M. Lai, W.R. Cullen, C.F. Harrington and K. Reimer, *Appl. Organomet. Chem.*, 1997, **11**, 797 .
39. Y. Shibata and M. Morita, *Appl. Organometal. Chem.*, 1992, **6**, 343.
40. J. Alberti, R. Rubio and G. Rauret, *Fresenius' J. Anal. Chem.*, 1995, **351**, 420.
41. S. Branch, L. Ebdon and P. O'Neill, *J. Anal. At. Spectrom.*, 1994, **9**, 33.
42. A.J.L. Muerer, A. Abildtrup, O.M. Poulsen and J.M. Christensen, *Analyst*, 1992, **117**, 677.
43. H. Helgesen and E.H. Larsen, *Analyst*, 1998, **123**, 791.
44. T. Dagnac, A. Padro, R. Rubio and G. Rauret, *Anal. Chim. Acta*, 1998, **364**, 19.
45. J.W. McKiernan, J.T. Creed, C.A. Brockhoff, J.A. Caruso and R.M. Lorenzana, *J. Anal. At. Spectrom.*, 1999, **14**, 607.
46. D. Heitkemper, J. Creed, J. Caruso and F.L. Fricke, *J. Anal. At. Spectrom.*, 1989, **4**, 279.
47. Y. Inoue, K. Kawabata, H. Takahashi and G. Endo, *J. Chromatogr A*, 1994, **675**, 149.
48. E.H. Larsen, G. Pritzl and S.H. Hansen, *J. Anal. At. Spectrom.*, 1993, **8**, 557.
49. X.R. Zhang, R. Cornelis, J. De Kimpe, L. Mees and N. Lameire, *Clin. Chem.*, 1997, **43**, 406.
50. M. Ma and X.C. Le, *Clin. Chem.*, 1998, **44**, 539.
51. S. McSheehy, F. Pannier, J. Szpunar, M. Potin-Gautier and R. Lobinski, *Analyst*, 2002, **127**, 223.
52. F. Lagarde, Z. Asfari, M.J.F. Leroy, C. Demesmay, M. Olle, A. Lamotte, P. Leperchec and E. A. Maier, *Fresenius' J. Anal. Chem.*, 1999, **363**, 12.
53. S. McSheehy, M. Marcinek, H. Chassaigne and J. Szpunar, *Anal. Chim. Acta*, 2000, **410**, 71.
54. M. Van Hulle, C. Zhang, X. Zhang and R. Cornelis, *Analyst*, 2002, **127**, 634.
55. K.A. Francesconi, S. Khokiattiwong, W. Goessler, S.N. Pedersen and M. Pavkov, *Chem. Commun.*, 2000, **12**, 1083.
56. K.A. Francesconi and J.S. Edmonds, *Rapid Commun. Mass Spectrom.*, 2001, **15**, 1641.
57. S. McSheehy, J. Szpunar, R. Lobinski, V. Haldys, J. Tortajada and J. Edmonds, *Anal. Chem.*, 2002, **74**, 2370.
58. R. Pickford, M. Miguens-Rodriguez, S. Afzaal, P. Speir, S.A. Pergantis and J.E. Thomas-Oates, *J. Anal. At. Spectrom.*, 2002, **17**, 173.

59. S.A. Pergantis, S. Wangkarn, K.A. Francesconi and J.E. Thomas-Oates, *Anal. Chem.*, 2000, **72**, 357 .

60. M. Miguens-Rodriguez, R. Pickford, J.E. Thomas-Oates and S.A. Pergantis, *Rapid Commun. Mass Spectrom.*, 2002, **16**, 323 .

61. B.R. Larsen, C. Astorga-Llorens, M.H. Florêncio and A.M. Bettencourt, *J. Chromatogr. A*, 2001, **926**, 167

CHAPTER 15

Speciation of Organoselenium Compounds in Biological Materials

1 Introduction

Selenium has been shown to be both essential for life and toxic at levels little above those required for health. Indeed, dietary levels of the desired amount of Se are in a very narrow range: consumption of food containing less than 0.1 mg kg^{-1} of this element will result in its deficiency whereas dietary levels above 1 mg kg^{-1} will lead to toxic manifestations.[1] Selenium exists in biological systems in the form of inorganic species such as Se(IV) (SeO$_3^{2-}$ - selenite), Se(VI) (SeO$_4^{2-}$ - selenate) or selenides (*e.g.* HgSe), or as organic species having a range of molecular masses and charges, starting from the simplest MeSeH and ending at complex selenoproteins.[2] Chemical forms of selenium reported in the context of speciation studies are summarised in Figure 15.1.

The essential nature of selenium results from its presence as a necessary component in the formation of the active center, the selenol group (-SeH), of glutathione peroxidase, thioredoxin reductase and of other selenoenzymes. Cancer chemopreventive effects are seen for inorganic selenium salts, selenoamino acids and various synthetic organoselenium compounds.[3] The metabolism of inorganic Se is complex and involves a number of species of which the most important are selenoamino acids: selenomethionine in microorganisms and plants, and seleno-cysteine in animals and man. In biological fluids selenium is bound to specific Se-binding proteins, and is a constituent of various proteins (containing covalent C-Se bonds of selenomethionine or selenocysteine).

The recognition of the protective role of selenium supplementation against cancer[4,5] and the wider availability of selenised yeast for this purpose[6] has raised concerns about the quality, safety and origin of marketed preparations. These issues can be addressed by appropriate analytical methodology allowing the

Selenite	SeO$_3^-$
Selenate	SeO$_4^-$
Trimethylselenonium ion	Me$_3$Se$^+$
Selenocysteine	H$_3$N$^+$-CH(COO$^-$)-CH$_2$-**SeH**
Selenocystine	H$_3$N$^+$-CH(COO$^-$)-CH$_2$-**Se-Se**-CH$_2$-CH(COO$^-$)-NH$_3^+$
Selenomethionine	H$_3$N$^+$-CH(COO$^-$)-CH$_2$-CH$_2$-**SeH**
Se- methylselenocysteine	H$_3$N$^+$-CH(COO$^-$)-CH$_2$-**Se**-CH$_3$
y-Glutamyl-Se-methylselenocysteine	H$_3$N$^+$-CH$_2$-CH$_2$-CO-NH-CH(COO$^-$)-CH$_2$-**Se**-CH$_3$
Selenocystathionine	H$_3$N$^+$-CH(COO$^-$)-CH$_2$-CH$_2$-**Se**-CH$_2$-CH(COO$^-$)-NH$_3^+$
Selenohomocysteine	H$_3$N$^+$-CH(COO$^-$)-CH$_2$-CH$_2$-**SeH**
Selenocystamine	H$_3$N$^+$-CH$_2$-CH$_2$-**Se-Se**-CH$_2$-CH$_2$-NH$_2$

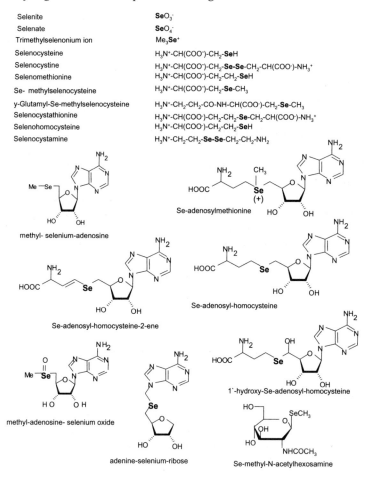

Figure 15.1 *Selenium species of interest in speciation analysis*

characterisation of selenised yeast in terms of the identity and concentrations of selenocompounds present.[7]

The areas of interest in the field of selenium speciation are multiple and each of them requires a dedicated analytical approach, especially in terms of the separation technique preceding the element- or molecule-specific detection. The myriad of selenium species with different physicochemical properties present in biological systems represents a challenge to the analyst. Since many selenium compounds have not yet been identified the role of molecular identification techniques such as GC MS for volatile species, ES MS/MS for selenopeptides and MALDI TOF for selenoproteins cannot be overestimated.

The major fields of interest include:

(1) speciation of volatile Se species in the environment.
(2) speciation of redox states in natural waters, soils and sediments.
(3) speciation of metabolic products (amino acids and peptides) in microorganisms, plants, and nutritional supplements.
(4) characterization of selenoproteins in yeast, plants and mammals.
(5) speciation of selenium metabolites in urine.

A particular field is the emerging need for the chiral separation of enantiomers of selenoamino acids.

Analytical techniques for the determination of selenium species in biological materials have been comprehensively reviewed[7-9] with a focus on selenoprotein analysis in mammals.[10] This chapter discusses analytical chemistry of selenium species related to biological materials. Speciation of volatile Se species in the environment is carried out by the purge-and-trap approaches discussed in Chapter 8. Speciation of redox states, Se(IV) and Se(VI), was discussed in Chapter 13.

2 Volatile Selenium Species in Plants

The volatile chemical species of Se are the reduced and methylated forms. The most well known are hydrogen selenide (H_2Se), methaneselenol (CH_3SeH), dimethylselenide ($(CH_3)_2Se$), dimethylselenylmethaneselenol (CH_3SeSCH_3), and dimethyl diselenide ($CH_3SeSeCH_3$). A few other less volatile mixed sulfur-selenium alkylated species have been detected in laboratory experiments with Se-rich plants such as garlic.[11] They include bis(methylthio)selenide ($CH_3SSeSCH_3$), methyl esters of propenesulfenoselenoic acids, and (allylthio)(methylthio)selenide ($CH_3SSeSCHCH=CH_2$).

The volatility of these species makes them amenable to gas chromatography. The detectors used have included MIP AES and ICP MS but the need for EI MS for species identification has been highlighted.[12] The relatively high boiling points of many of these species require purging from saturated Na_2SO_4 solution at elevated temperatures (90 °C). The headspace is injected into capillary GC using a temperature programmed oven and interface heated efficiently to avoid condensation on the way to MIP AED detector.[12]

Using capillary GC-MIP AED a number of organoselenium compounds have been detected in the headspace over allium plants, *e.g.* elephant garlic.[12,13] The identity of the compounds can be determined by GC-MS or by comparison of retention times to those of chemical standards. A chromatogram is shown in Figure 15.2. The species detected are usually mixed selenium/sulfur compounds.

Some non-volatile selenium species can be derivatised for analysis by GC. The common techniques for amino acid derivatisation have been successfully used.[11] The methods are rather tedious and may fail for oligopeptides. Therefore it is preferable to determine selenoamino acids and selenooligopeptides using HPLC coupled with ICP MS.

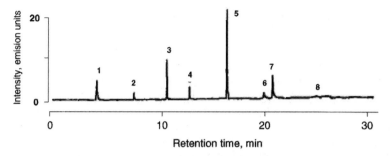

Figure 15.2 *An MIP AES chromatogram of volatile selenocompounds found in the head-space of a garlic extract: 1 – dimethylselenide, 2 – allyl methyl selenide, 3 – methanesulfenoselenoic acid methyl ester, 4 – dimethyl diselenide, 5 – 2-propenesulfenoselenoic acid methyl ester, 6 – 1-propenesulfenoselenoic acid methyl ester, 7 – bis(methylthio)selenide, 8 – (allylthio)(methylthio)selenide* (Reprinted from *J. Chromatogr. A*, 1995, **703**, 393, copyright 1995, with permission from Elsevier)

3 Selenoamino Acids and Selenopeptides in Yeast and Plants

A number of selenoamino acids have been identified in microorganisms and plants, besides the most common selenate and selenite species. In the most widely investigated sample, Se-enriched yeast, more than 20 Se compounds including selenocysteine, selenomethionine, methylselenocysteine, and inorganic forms seem to be present.[14–18] Five selenium species and several unknown peaks have been observed in selenium-enriched garlic, onion and broccoli.[19,20] Several seleno-analogues of sulfur-containing amino acids and their derivatives have been identified in terrestrial plants, especially in so-called 'selenium accumulator plants'.[21–23] Among the six organoselenium compounds detected in seagull eggs the major compounds are selenocysteine and selenocystamine.[24] Another area of interest in species-selective determination of amino acids is their formation as the product of enzymolysis of selenoproteins.[25–28]

Sample Preparation Procedures

Since selenoamino acids are water-soluble, leaching with hot water has been judged sufficient to recover selenium species not incorporated in larger molecules. The sample is homogenised with water, sonicated or heated and ultracentrifuged. The typical recovery of selenium extracted in this way is *ca.* 10–15%,[14,15,17,25,29] but it can be total in the case of yeast samples enriched by doping with selenite. Selenoamino acids have been separated by ultrafiltration (breast milk)[30] or dialysis (algal extract).[26] Selenocysteine and some other selenoamino acids are highly susceptible to oxidative degradation, because the selenol group has a significantly lower oxidation potential than its sulfur counterpart. The carboxymethyl derivative can be synthesized (by addition of iodoacetic acid) to stabilise selenocysteine and thus prevent its degradation.[31]

The low yields of the aqueous leaching procedure for some species and samples has promoted the use of more aggressive leaching media by some researchers. A trade-off is always necessary between the recovery of Se from a solid matrix and the preservation of the original Se species. As shown by Casiot *et al.*,[17] the addition of SDS to the leaching mixture increases the yield of Se by releasing selenoproteins into the aqueous phase. The recovery of selenoamino acids can increase to above 95% by degrading the species originally present with a mixture of proteolytic enzymes.[25] Low extraction efficiencies have been obtained for bacterial samples (1% hot water, 8% protease and HCl hydrolysis, 12% lysozyme and lysozyme-protease).[32] Figure 15.3 shows typical recoveries from selenised yeast and size exclusion chromatographic profiles of extracts obtained with different extractant solutions.

Care is advised in the interpretation of literature data since the results depend on the way the sample was prepared. This applies in particular to the frequently used statement 'the majority of Se is present as selenomethionine', when describing the result of a procedure involving an enzymic digestion. Actually, selenomethionine usually constitutes a part of a larger stable selenium-containing protein that has been destroyed during the sample preparation procedure.

Determination of Selenoamino Acids

A plethora of methods claimed to be useful for the speciation of Se(IV), Se(VI), selenomethionine and selenocysteine exist, as can be seen from the review papers.[33-35] The major reason for this choice of analytes seems to be the commercial availability of standards for these compounds since many of the methods developed have never been applied to real life samples. Paradoxically, the opposite approach, *i.e.* one based on the screening of a sample for the presence of stable selenium species by HPLC with Se-specific detection, leads not only to the conclusion that the actual number of species present in selenised yeast exceeds 20 but also that none of the four above mentioned standard compounds is apparently the dominant species.[14-18]

The direct coupling of HPLC to ICP-MS is an established approach to the determination of Se species in biological materials. Employed separation mechanisms have included anion and cation exchange and cation pairing reversed phase chromatography. The latter (using perfluorinated carboxylic acids as cation pairing reagents) offers the highest resolution and should be given preference for the characterisation of samples containing many selenocompounds.[21,36] A typical chromatogram obtained for a water extract of an yeast sample is shown in Figure 15.4. Reversed phase HPLC-ICP MS has also been demonstrated to be well suited for the separation of selenopeptides resulting from tryptic digestion of selenoproteins.[37]

In cases where interest is limited to the quantification of inorganic selenium species or selenomethionine (enzymic extracts) anion exchange HPLC seems to be the technique of choice. The chromatographic mobile phase should possess a pH buffer effect in the range of the pK_a values of the anionic Se compounds and,

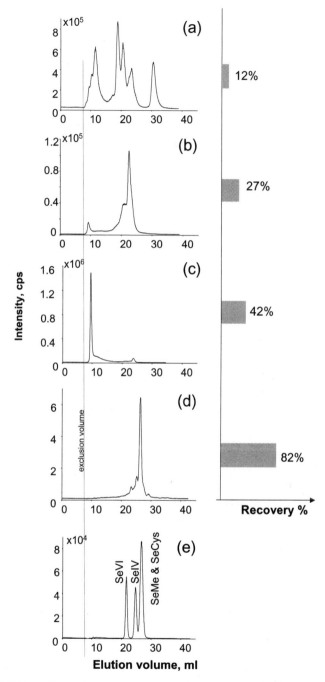

Figure 15.3 *Effect of sample preparation on the recovery of selenium species from selenised yeast by SEC-ICP MS: (a) leaching with hot H₂O; (b) pectinolysis with Driselase; (c) leaching with SDS solution, (d) enzymatic proteolysis, (e) standards*

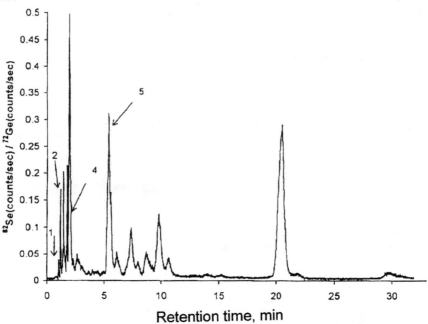

Figure 15.4 *An ion pairing reversed phase ICP MS chromatogram for a comprehensive screening of plant extracts for organoselenium species: 1 – selenate, 2 – selenite, 4 – Se-methyl-DL-selenocysteine, 5 – DL-selenomethionine*
(Reprinted from *Spectrochim. Acta.*, 2001, **57B**, 173, copyright 2001, with permission from Elsevier)

in order to prevent salt build-up on the sampler and skimmer cones, the eluting molecular species should preferably be organic in nature.[38] An application of CZE-ICP MS has been shown for the analysis of selenoamino acids in breast milk[30,39] and selenised yeast.[40]

Retention time matching with commercially available standards has commonly been used for the identification of selenium species, usually with little success. The alternatives include the synthesis of a significant number of selenium compounds expected to be found in the analysed samples[21,36] or the use of ES MS[20,27] or ES MS/MS[18] for the identification of Se species eluted from an HPLC column.

Multidimensional Total Characterisation Approaches

An analytical protocol aimed at the identification of selenocompounds in a biological sample is shown schematically in Figure 15.5. It is based on the purification of the compound(s) of interest by multidimensional chromatography followed by the characterisation of isolated compounds by tandem mass spectrometry.

SEC-ICP MS offers sufficiently high resolution to allow the fractionation of an aqueous yeast extract into 3–6 fractions (Figure 15.5a). Each of them is further

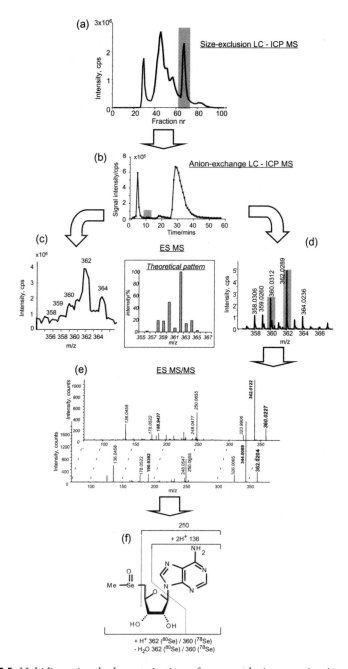

Figure 15.5 *Multidimensional characterisation of organoselenium species in yeast and plant materials*
(reproduced with permission from *Fresenius' J. Anal. Chem.*, 2002, **370**, 597, copyright 2002, Springer-Verlag)

fractionated by anion exchange chromatography according to the electrical charge or polarity (Figure 15.5b). Cationic or uncharged compounds elute as a mixture in the void or close to it and need to be further purified by cation exchange HPLC. The fractions obtained, after 2-D HPLC for anionic compounds and after 3-D HPLC for other compounds, are analysed by electrospray MS, using a quadrupole (Figure 15.5c) or a TOF mass analyser (Figure 15.5d). Note that such an extensive multidimensional separation is necessary not to separate the selenocompounds from each other but rather to remove the matrix components which would otherwise affect the electrospray ionisation. The choice and optimisation of volatile buffers which do not suppress the electrospray ionisation is mandatory.

The characteristic isotopic pattern of Se should allow the recognition of peaks of Se species in the mass spectrum. In practice, however, the sensitivity of a quadrupole mass analyser is often insufficient to detect the selenium species and, even for the detected compounds, the recognition of the Se isotopic pattern may be far from certain (Figure 15.5c). Therefore the use of ES TOF MS is recommended. Not only the sensitivity is higher, but also the resolution of 10 000 allows the clear identification of a selenium pattern (Figure 15.5d). Owing to the high mass measurement accuracy a hypothesis concerning the empirical formula of the analyte compound can be put forward facilitating its identification.

The structural characterisation of an isolated compound is further possible by collision induced dissociation of a protonated molecular ion selected by a quadrupole mass filter. In the case of selenium, protonated ions corresponding to the two most abundant Se isotopes ^{78}Se and ^{80}Se are usually fragmented. This allows easy discrimination in the CID mass spectrum of peaks from fragments that contain Se (the m/z signals in both spectra appear at a difference of 2 u) and fragments that do not contain Se (the m/z signals in both spectra have the same value), which helps in the interpretation of mass spectra. The use of a TOF mass analyser at the second stage of tandem MS has the benefit of rapidly scanning the whole mass range which provides enough time for the fragmentation of more molecular ions (on the basis of the same sample volume available) and often the detection of further Se compounds hidden in the background noise of the ES MS. Figure 15.5e shows CID mass spectra of a recently discovered selenocompound with an M_r value (for the species containing ^{80}Se isotope) of 362.0289 which corresponds to the empirical formula $C_{11}H_{15}N_5O_4Se$ (theoretical 362.0364, $\delta = -20$ ppm) . The tandem mass spectra allow the recognition of Se-containing fragments and Se-free fragments (highlighted in bold font) and the assignment of a structural formula (Figure 5f) to the compound.[42]

The above outlined (or simplified) analytical strategy has been successfully applied to the identification of a number of selenocompounds in yeast,[18,41-43] garlic[44] and selenium accumulating plants.[23,45]

4 Selenoproteins

In mammals, speciation of Se usually involves the determination of the different Se-containing proteins.[46-53] The most important seem to be selenoprotein P, a major protein which is sometimes used as a biochemical marker of selenium status,[47] and selenoenzymes such as several glutathione peroxidases and type 1 iodothyronine de-iodinase.[46,47,51] More than 25 Se-containing proteins or protein sub-units have been detected in rat tissues labelled *in vivo* with [75]Se.[51] Two types of analytical approach have been developed for the identification and determination of selenoproteins: one based on LC (size exclusion and affinity) and the other on flat-bed electrophoresis.

Despite the fact that SE HPLC was the first to produce evidence of the presence of selenoprotein P in human plasma, it is considered to lack resolution and cannot separate the major Se proteins. Also, the large dilution factor limits the detection sensitivity. SE HPLC-ICP MS of a human serum sample yields three signals, none of which, however, co-elute with the glutathione peroxidase activity.[52] SE HPLC of human breast milk whey produces four Se signals corresponding to species with apparent molecular weights of 15, 60, 1500 and >2000 kDa.[54]

The principal speciation approach for selenoproteins in serum is based on affinity chromatography; column packing materials are available that allow the separation of serum Se proteins.[46,49,50] Affinity chromatography using a heparin-Sepharose CL-6B column in series with a column of reactive blue 2-Sepharose CL-6B with off-line detection has been proposed for the separation of selenoproteins in serum.[46,49] Recently, a combination of affinity chromatography with SE HPLC with on-line ICP MS detection has been reported to separate the three major Se-containing proteins (albumin, glutathione peroxidase and selenoprotein P) found in human plasma.[53]

An enriched stable isotope ([82]Se) has been used to study the fate of injected and endogenous Se by HPLC-ICP MS.[55-57] The unambiguous identification of Se peaks was not possible; the identity of some of them was postulated on the basis of the molecular size estimated from the SE HPLC chromatograms.[55-58]

Flat-bed electrophoresis offers much better resolution than HPLC for the separation of selenoproteins. The fact that Se is bound *via* a covalent bond allows the use of SDS PAGE without the risk of Se losses.[10,51,59-64] The principal approaches are based on *in vivo* labelling of rats with [75]Se (by injection of the tracer to Se-depleted animals) which was then detected in the gel strips by autoradiography[10,51,59,61,62,64] and INAA.[60,61,65] The sensitivity of the techniques is 0.02 pg and 0.1 ng, respectively.[66] A few tens of different selenoproteins have been discussed in a review paper[51] and in the above cited work. Differences between tissues have been observed indicating specific functions of some of the Se-containing proteins. Differences have also been observed between the subcellular fractions indicating that some Se proteins are involved in different intracellular processes.[59] SDS PAGE has also been used for the detection of selenoprotein in soybean radiolabelled with [75]Se.[65,67] Prior to PAGE the isolation of the protein fraction by anion-exchange chromatography is recommended.[67]

Two-dimensional high resolution gel electrophoresis has been applied to the fractionation of ^{75}Se-containing proteins in yeast (grown in ^{75}Se-containing medium).[68] Using autoradiography detection, a preliminary 2-D map of selenium-containing proteins in yeast has been established.[68]

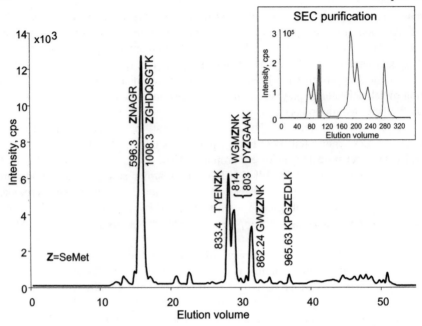

Figure 15.6 *Characterisation of an enzymic digest of water-soluble selenium-containing proteins (fraction purified by SEC as shown in the inset) from selenised yeast by reversed phase HPLC with ICP MS detection*

MALDI-TOF MS has been proposed as a key technique in a novel generic approach to speciation analysis of selenium in yeast supplements.[37] Owing to a lower detection limit and superior matrix tolerance to electrospray MS it has allowed the successful detection of selenocompounds in samples for which electrospray MS has failed. The analytical approach developed has been applied to the identification of previously unreported selenopeptides in the tryptic digest of a water-soluble selenoprotein fraction isolated by size exclusion chromatography and purified by RP chromatography. The information on the mass of the protonated molecular ion obtained from MALDI has allowed the optimisation of the conditions for collision induced dissociation MS using a triple quadrupole spectrometer thus enabling the determination of the amino acid sequences.[37] Figure 15.6 shows a reversed phase HPLC-ICP MS chromatogram of an enzymic digest of an extract of water-soluble proteins from yeast, annotated with the molecular masses and peptides sequences determined by MALDI-TOF MS and ES MS/MS.

5 Selenium Metabolites in Urine

The reduction of Se followed by its methylation to methylselenol, dimethylselenide and the trimethylselenonium ion is the primary pathway for the metabolism of selenite by mammals. Some of these and other unidentified compounds of this type are excreted in urine.[57,69] In view of the low concentrations involved, the use of an ultrasonic nebuliser to interface HPLC with ICP MS,[70] a hydraulic high pressure nebuliser with ICP collision cell MS[71] or the use of a HR ICP mass spectrometer[72] are advised. Detection limits at the sub $ng\,mL^{-1}$ level can be obtained.

Trimethylselenonium accounts for a few percent of Se in urine in normal subjects but it usually becomes predominant if the nutritional intake of selenium increases, probably due to detoxification in the kidney. Monomethylselenium has been found in untreated rats but rats to which selenate had been administered by injection excreted trimethylselenonium.[72] Three signals have been observed in RP HPLC chromatograms of basal human urine but their identification by retention time matching is unconvincing.[72] At least five species have been found in urine but only two organoselenium compounds, trimethylselenonium and monomethylselenol, have been identified.[57] Six distinct selenium-containing species have been detected in human urine by reversed phase HPLC-ICP MS, the first two being identified off-line by ES MS/MS as selenomethionine and selenocystamine.[73] A typical HPLC-ICP MS urine chromatogram is shown in Figure 15.7.[74]

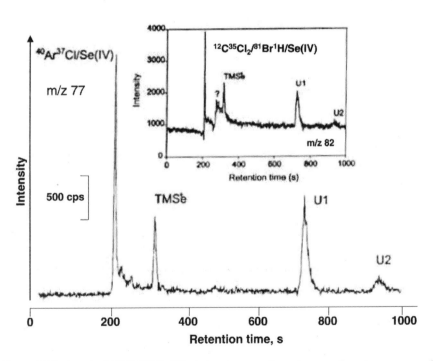

Figure 15.7 *A typical HPLC-ICP MS chromatogram of a urine sample using a reversed phase separation mechanism with mixed ion pair reagents; a corresponding chromatogram at the m/z 82 is shown in the inset (U1, U2 – unknown)* (from Ref. 74)

Ogra *et al.*[75] have identified a novel selenium metabolite in rat urine – the selenosugar diastereoisomer *Se*-methyl-*N*-acetylhexosamine – by ES MS/MS after purification consisting of removal of urea and sodium chloride. The finding has been confirmed in human urine.[76] Extraction of urine samples with benzo-15-crown-5-ether to remove sodium and potassium results in the improvement of the chromatographic separation of selenium compounds present by anion exchange HPLC-ICP MS.[77] Selenomethionine and trimethylselenonium have been identified on the basis of their retention times out of the 6–7 peaks present.[77]

6 Optically Active Selenospecies

Selenomethionine and other α-selenoamino acids and related compounds are chiral and the different enantiomers have different physiological activity. A variety of techniques including GC, HPLC, CE and micellar electrokinetic chromatography, usually coupled with ICP MS, have been proposed for their speciation analysis.

A Chirobiotic T column is preferred to a β-cyclodextrin column for the chiral separation of selenoamino acids by HPLC.[78] Derivatisation with *o*-phthalaldehyde or naphthalene-2,3-dicarboxaldehyde is required.[79] HPLC-ICP MS with chiral crown ether stationary phase has been applied to speciation of D and L forms of selenomethionine in yeast nutritional supplements.[80]

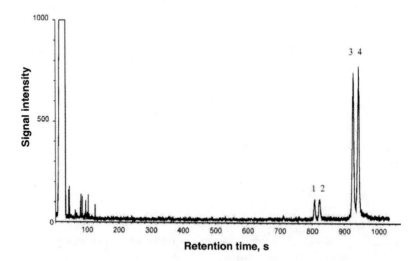

Figure 15.8 *Speciation of optically active selenoamino acids by chiral capillary GC-ICP MS (standard mixture of the racemates of D,L-selenomethionine at 25 mg ml⁻¹ and D,L-selenoethionine at 250 mg ml⁻¹ as the N-ethoxycarbonyl-O-ethyl ester derivatives on Chirasil-L-Val): 1 – D-selenomethionine, 2 – L-selenomethionine, 3 – D-selenoethionine, 4 – L-selenoethionine*
(Reproduced from *J. Pharm. Biomed. Anal.*, 2002, **27**, 507, copyright 2002, with permission from Elsevier)

Capillary electrophoresis has been proposed to separate the enantiomers of selenoamino acids in the presence of vancomycin or β-cyclodextrin.[81] Enantiomeric separation of selenoamino acid derivatives by β-cyclodextrin-modified micellar electrokinetic chromatography using a mixed micellar system of SDS and taurodeoxycholic acid and prior derivatisation with 2,3-naphthalenedicarboxaldehyde to produce cyanobenzoisoindole derivatives has been reported.[82]

Gas chromatography using a Chirasil-L-Val column and ICP MS detection has allowed a detection limit of 4 pg to be obtained. The enantiomers of selenomethionine were converted into their *N*-fluoroacetyl (TFA)-*O*-alkyl esters.[78,83] A typical GC-ICP MS chromatogram of selenoamino acids enantiomers is shown in Figure 15.8.

References

1. L.H. Foster and S. Sumar, *Crit. Rev. Food Sci. Nutr.*, 1997, **37**, 211.
2. Y. Shibata, M. Morita and K. Fuwa, *Adv. Biophys.*, 1992, **28**, 31.
3. G.F.J. Combs, L.C. Clark and B.W. Turnbull, *Biofactors*, 2001, **14**, 153.
4. L.C. Clark, J.G. Combs, B.W. Turnbill, E.H. Slate, D.K. Chalker, J. Chow, K.S. Davis, R.A. Glover, G.F. Graham, E.G. Gross, A. Krongrad, J.L. Lesher, K. Park, B.B. Sanders, C.L. Smith and J.R. Taylor, *J. Am. Med. Assoc.*, 1996, **276**, 1957.
5. M.P. Rayman, *Lancet*, 2000, **356**, 233.
6. J. Neve, in P. Collery, P. Bratter, V. Negrettii de Bratter, I. Khassanova and J.C. Etienne (eds), *Metal Ions in Biology and Medicine,* John Libbey, Paris, 1998.
7. R. Lobinski, J.S. Edmonds, K.T. Suzuki and P.C. Uden, *Pure Appl. Chem.*, 2000, **72**, 447.
8. K. Pyrzynska, *Analyst*, 1996, **121**, 77R.
9. P.C. Uden, *Anal. Bioanal. Chem.*, 2002, **373**, 422.
10. D. Behne, C. Hammel, H. Pfeifer, D. Rothlein, H. Gessner and A. Kyriakopoulos, *Analyst*, 1998, **123**, 871.
11. X.J. Cai, E. Block, P.C. Uden, X. Zhang, B.D. Quimby and J.J. Sullivan, *J. Agric. Food Chem.*, 1995, **43**, 1754.
12. E. Block, X.J. Cai, P.C. Uden, X. Zhang, B.D. Quimby and J.J. Sullivan, *Pure Appl. Chem.*, 1996, **68**, 937.
13. P.C. Uden, *J. Chromatogr. A*, 1995, **703**, 393.
14. S.M. Bird, H. Ge, P.C. Uden, J.F. Tyson, E. Block and E. Denoyer, *J. Chromatogr. A*, 1997, **789**, 349.
15. S.M. Bird, P.C. Uden, J.F. Tyson, E. Block and E. Denoyer, *J. Anal. At. Spectrom.*, 1997, **12**, 785.
16. P.C. Uden, S.M. Bird, M. Kotrebai, P. Nolibos, J.F. Tyson, E. Block and E. Denoyer, *Fresenius' J. Anal. Chem.*, 1998, **362**, 447.
17. C. Casiot, J. Szpunar, R. Lobinski and M. Potin Gautier, *J. Anal. At. Spectrom.*, 1999, **14**, 645.

18. C. Casiot, V. Vacchina, H. Chassaigne, J. Szpunar, M. Potin-Gautier and R. Lobinski, *Anal. Commun.*, 1999, **36**, 77.
19. H. Ge, X.J. Cai, J.F. Tyson, P.C. Uden, E.R. Denoyer and E. Block, *Anal. Commun.*, 1996, **33**, 279.
20. M. Kotrebai, M. Biringer, J.F. Tyson, E. Block and P.C. Uden, *Anal. Commun.*, 1999, **36**, 249.
21. M. Kotrebai, S.M. Bird, J.F. Tyson, E. Block and P.C. Uden, *Spectrochim. Acta*, 1999, **54B**, 1573.
22. M. Montes-Bayon, E.G. Yanes, C. Ponce de Leon, K. Jayasimhulu, A. Stalcup, J. Shann and J.A. Caruso, *Anal. Chem.*, 2002, **74**, 107.
23. M. Montes-Bayon, D.L. LeDuc, N. Terry and J.C. Caruso, *J. Anal. At. Spectrom.*, 2002, **17**, 872.
24. N. Jakubowski, C. Thomas, D. Klueppel and D. Stuewer, *Analysis*, 1998, **26**, M37.
25. N. Gilon, A. Astruc, M. Astruc and M. Potin-Gautier, *Appl. Organomet. Chem.*, 1995, **9**, 623.
26. T.M. Fan, A.N. Lane, D. Martens and R.M. Higashi, *Analyst*, 1998, **123**, 875.
27. H.M. Crews, P.A. Clarke, D.J. Lewis, L.M. Owen, P.R. Strutt and A. Izquierdo, *J. Anal. At. Spectrom.*, 1996, **11**, 1177.
28. M.A. Beilstein, P.D. Whanger and G.Q. Yang, *Biomed. Environ. Sci.*, 1991, **4**, 392.
29. J. Zheng, W. Goessler and W. Kosmus, *Trace Elements Electrol.*, 1998, **15**, 70.
30. B. Michalke and P. Schramel, *J. Chromatogr. A*, 1998, **807**, 71.
31. K. Takatera, N. Osaki, H. Yamaguchi and T. Watanabe, *Anal. Sci.*, 1994, **10**, 567.
32. B. Michalke, H. Witte and P. Schramel, *Anal. Bioanal. Chem.*, 2002, **372**, 444.
33. X. Dauchy, M. Potin Gautier, A. Astruc and M. Astruc, *Fresenius' J. Anal. Chem.*, 1994, **348**, 792.
34. K. Pyrzynska, *Chem. Anal.*, 1995, **40**, 677.
35. G. Kölbl, K. Kalcher, K.J. Irgolic and R.J. Magee, *Appl. Organomet. Chem.*, 1992, **7**, 443.
36. M. Kotrebai, J.F. Tyson, E. Block and P.C. Uden, *J. Chromatogr. A*, 2000, **866**, 51.
37. J. Ruiz Encinar, R. Ruzik, W. Buchmann, J. Tortajada, R. Lobinski and J. Szpunar, *Analyst*, 2003, **128**, 220.
38. G. Alsing Pedersen and E.H. Larsen, *Fresenius' J. Anal. Chem.*, 1997, **358**, 591.
39. B. Michalke, *J. Chromatogr. A*, 1995, **716**, 323.
40. S. Mounicou, S. McSheehy, J. Szpunar, M. Potin-Gautier and R. Lobinski, *J. Anal. At. Spectrom.*, 2002, **17**, 15.
41. S. McSheehy, V. Haldys, J. Tortajada and J. Szpunar, *J. Anal. At. Spectrom.*, 2002, **17**, 507.
42. J. Szpunar and R. Lobinski, *Anal. Bioanal. Chem.*, 2002, **373**, 404.

43. S. McSheehy, F. Pannier, J. Szpunar, M. Potin-Gautier and R. Lobinski, *Analyst*, 2002, **127**, 223.
44. S. McSheehy, W. Yang, F. Pannier, J. Szpunar, R. Lobinski, J. Auger and M. Potin-Gautier, *Anal. Chim. Acta*, 2000, **421**, 157.
45. M. Montes-Bayon, T.D. Grant, J. Meija and J.A. Caruso, *J. Anal. At. Spectrom.*, 2002, **17**, 105.
46. I. Harrison, D. Littlejohn and G.S. Fell, *Analyst*, 1996, **121**, 189.
47. M. Persson Moschos, W. Huang, T.S. Srikumar, B. Akesson and S. Lindeberg, *Analyst*, 1995, **120**, 833.
48. C.D. Thomson, *Analyst*, 1998, **123**, 827.
49. J.T. Deagen, J.A. Butler, B.A. Zachara and P.D. Whanger, *Anal. Biochem.*, 1993, **208**, 176.
50. T. Plecko, S. Nordmann, M. Rükgauer and J.D. Kruse-Jarres, *Fresenius' J. Anal. Chem.*, 1999, **363**, 517.
51. D. Behne, C. Weiss Nowak, M. Kalcklosch, C. Westphal, H. Gessner and A. Kyriakopoulos, *Analyst*, 1995, **120**, 823.
52. H. Koyama, Y. Kasanuma, C. Kim, A. Ejima, C. Watanabe, H. Nakatsuka and H. Satoh, *Tohoku J. Exp. Med.*, 1996, **178**, 17.
53. H. Koyama, K. Omura, A. Ejima, Y. Kasanuma, C. Watanabe and H. Satoh, *Anal. Biochem.*, 1999, **267**, 84.
54. V.E. Negretti de Braetter, S. Recknagel and D. Gawlik, *Fresenius' J. Anal. Chem.*, 1995, **353**, 137.
55. Y. Shiobara and K.T. Suzuki, *J. Chromatogr. B: Biomed Appl.*, 1998, **710**, 49.
56. K.T. Suzuki, M. Itoh and M. Ohmichi, *J. Chromatogr. B, Biomed Appl.*, 1995, **666**, 13.
57. K.T. Suzuki, M. Itoh and M. Ohmichi, *Toxicology*, 1995, **103**, 157.
58. K.T. Suzuki, *Tohoku J. Exp. Med.*, 1996, **178**, 27.
59. D. Behne, H. Hilmert, S. Scheid, H. Gessner and W. Elger, *Biochim. Biophys. Acta*, 1988, **966**, 12.
60. D. Behne, C. Weiss-Nowak, M. Kalcklosch, C. Westphal, H. Gessner and A. Kyriakopoulos, *Biol. Trace Elem. Res.*, 1994, **43-45**, 287.
61. D. Behne, S. Scheid, H. Hilmert, H. Gessner, D. Gawlik and A. Kyriakopoulos, *Biol. Trace Elem. Res.*, 1990, **26-27**, 439.
62. D. Behne, A. Kyriakopoulos, M. Kalckosch, C. Weiss-Nowak, H. Pfeifer, H. Gessner and C. Hammel, *Biomed. Environ. Sci.*, 1997, **10**, 340.
63. C. Hammel, A. Kyriakopoulos, U. Roesick and D. Behne, *Analyst*, 1997, **122**, 1359.
64. A. Kyriakopoulos, M. Kalcklosch, C. Weiss-Nowak and D. Behne, *Electrophoresis*, 1993, **14**, 108.
65. C. Hammel, A. Kyriakopoulos, D. Behne, D. Gawlik and P. Brätter, *J. Trace Elem. Med. Biol.*, 1996, **10**, 96.
66. S.F. Stone, G. Bernasconi, N. Haselberger, M. Makarewicz, R. Ogris, R. Wobrauschek and R. Zeisler, *Biol. Trace Elem. Res.*, 1994, **43-45**, 299.
67. S.K. Sathe, A.C. Mason, R. Rodibaygh and C.M. Weaver, *J. Agric. Food Chem.*, 1992, **40**, 2084.

68. C.C. Chery, E. Dumont, R. Cornelis and L. Moens, *Fresenius' J. Anal. Chem.*, 2001, **371**, 775.
69. Y. Shiobara, Y. Ogra and K.T. Suzuki, *Analyst*, 1999, **124**, 1237.
70. K.L. Yang and S.J. Jiang, *Anal. Chim. Acta*, 1995, **307**, 109.
71. J.M. Marchante-Gayon, I. Feldmann, C. Thomas and N. Jakubowski, *J. Anal. At. Spectrom.*, 2001, **16**, 457.
72. J.M. Gonzalez LaFuente, J.M. Marchante-Gayon, M.L. Fernandez Sanchez and A. Sanz-Medel, *Talanta*, 1999, **50**, 207.
73. T.H. Cao, R.A. Cooney, M.M. Woznichak, S.W. May and R.F. Browner, *Anal. Chem.*, 2001, **73**, 2898.
74. J. Zheng, M. Ohata and N. Furuta, *J. Anal. At. Spectrom.*, 2002, **17**, 730.
75. Y. Ogra, K. Ishiwata, H. Takayama, N. Aimi and K.T. Suzuki, *J. Chromatogr. B*, 2002, **767**, 301.
76. B. Gammelgaard, K.G. Madsen, J. Bjerrum, L. Bendahl, O. Jons, J. Olsen and U. Sidenius, *J. Anal. At. Spectrom.*, 2003, **18**, 65.
77. B. Gammelgaard, O. Joens and L. Bendahl, *J. Anal. At. Spectrom.*, 2001, **16**, 339.
78. S. Perez Mendez, E. Blanco Gonzalez and A. Sanz-Medel, *Biomed. Chromatogr.*, 2001, **15**, 181.
79. S. Perez Mendez, E. Blanco Gonzalez, M.L. Fernandez Sanchez and A. Sanz-Medel, *J. Anal. At. Spectrom.*, 1998, **13**, 893.
80. K. Sutton, C.A. Ponce de Leon, K.L. Ackley, R.M.C. Sutton, A.M. Stalcup and J.A. Caruso, *Analyst*, 2000, **125**, 281.
81. K.L. Sutton, R.M. C. Sutton, A.M. Stalcup and J.A. Caruso, *Analyst*, 2000, **125**, 231.
82. S. Perez Mendez, E. Blanco Gonzalez and A. Sanz-Medel, *Anal. Chim. Acta*, 2000, **416**, 1.
83. S. Perez Mendez, M. Montes Bayon, E. Blanco Gonzalez and A. Sanz-Medel, *J. Anal. At. Spectrom.*, 1999, **14**, 1333.
84. C. Devos, K. Sandra and P. Sandra, *J. Pharm. Biomed. Anal.*, 2002, **27**, 507.

CHAPTER 16

Speciation of Metal Complexes in Microorganisms, Plants and Food of Plant Origin

1 Introduction

Essential metals in small quantities are required by microorganisms and plants to accomplish specific catalytic functions. Toxic elements, and essential elements at higher levels, disturb the metabolism which results in the deactivation of essential enzymatic reactions, damage to membranes and mutagenesis. Microorganisms and plants have developed a number of internal mechanisms to control the homeostasis of essential elements and to cope with the stress induced by toxic elements.[1] Some plants, referred to as hyperaccumulators, have evolved particularly efficient metal homeostasis mechanisms developed which allow them to live and reproduce in metal-rich environments.[2]

Resistance mechanisms include the high turnover of organic acids such as phytate, malate, citrate, oxalate, succinate or others and the induction and activation of antioxidant enzymes, such as superoxide dismutase or glutathione peroxidase.[1] A well known mechanism for enhancing heavy metal accumulation and tolerance is the expression of metal-binding proteins or peptides in plants.[3] Metals complexed by uronic acid derivatives are components of cell walls complexed by pectin subunits.[4] The complexation of metals leads to a number of relatively poorly characterised metal complexes. The understanding of the mechanisms controlling detoxification is limited only by the availability of analytical data on the species formed.

Various analytical approaches have been proposed to study metal speciation in plants.[5] Except for a few, their common feature is the analyte recovery procedure based on the sample homogenization with an aqueous buffer, sonication, ultracentrifugation and the analysis of the supernatant. Works to examine trace element speciation in the solid residue have been scarce. Extraction techniques lead to the homogenisation of cell compartments and cells of different types thus leading to potential ligand exchange. The ultimate goal is therefore analysis of a single cell type or of the subcellular compartments.

Size exclusion LC-ICP MS is a convenient technique for screening for the presence of stable metal complexes with amino acids, polysaccharides or proteins within plants.[6] The use of other separation mechanisms has been scarce probably due to difficulties with preservation of metal-ligand complexes in other than physiological conditions.[7] Complementary data can be obtained by demetallation of complexes followed by reversed phase HPLC or capillary electrophoresis of the ligands present in fractions heart-cut from size exclusion LC. Electrospray MS is an essential technique for species identification.[8]

2 Metal Complexes with Water-Soluble Proteins and Polypeptides

Phytochelatins

Phytochelatins (PCs) are a class of oligopeptides composed of only three amino acids: cysteine (Cys), glutamic acid (Glu) and glycine (Gly) and in which glutamic acid is linked to cysteine through a γ-peptide linkage. Their general formula is $(GluCys)_n Gly$ where n is between 2 and 11.[9,10] PCs can detoxify metals by forming a metal-PC complex in which the metal is bound to the thiol group of the cysteine unit. The general structure of phytochelatins is conservative in a wide variety of plants but some modifications may occur on the C-terminal amino acid leading to the loss of glycine or its replacement by β-alanine, serine, glutamine or glutamic acid.[9,10] Some microorganisms (cyanobacterium) have been reported to bioinduce higher metal-complexing polypeptides referred to as metallothionein-like proteins.[11–14] Structural formulae of the most commonly found PC families are shown in Figure 16.1.

Size exclusion LC-ICP MS has been proposed as a convenient technique for the monitoring of Cd, Cu and Zn complexes in aqueous extracts of microorganisms and plants that have been submitted to metal stress.[11–13,15–18] Under carefully optimised conditions this technique allows the separation of complexes with PC ligands having different lengths of peptide chain.[17] A typical SEC-ICP MS chromatogram for a mixture of standards is shown in Figure 16.2. However, the formation of mixed-ligand complexes is common which has resulted in an (overlapped) chromatogram of a sample (soybean extract) showing a phytochelatin peak of which the elution volume does not match any of the standards available. Therefore, information on the bioligands induced as a response to metal stress can be obtained only after heart-cutting of the metal-containing fraction and its characterisation by a different separation technique, such as CZE-ES MS/MS.[19] A total ion current chromatogram with a mass spectrum extracted at the peak apex, annotated with amino acid sequences determined by tandem MS is shown in Figure 16.3.

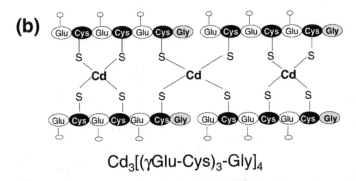

(a) $(\gamma\text{–Glu-Cys})_n\text{-X}$ (n=2-11)
X = no, Gly, Glu, Gln, Ser, β-Ala

(b)

$Cd_3[(\gamma Glu\text{-}Cys)_3\text{-}Gly]_4$

Figure 16.1 *Structure of the most common phytochelatin families: (a) general formula; (b) example of the $Cd_3[(\gamma\text{-}Glu\text{-}Cys)_3\text{-}Gly]_4$ complex*

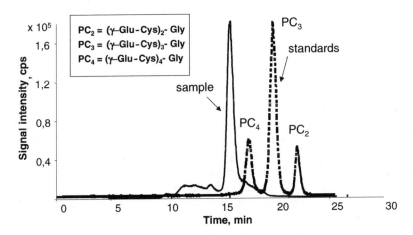

Figure 16.2 *Size exclusion chromatography – ICP MS of a plant (soybean) extract containing phytochelatins. A chromatogram of a mixture of PC_2, PC_3 and PC_4 standards acquired in identical conditions is overlapped. Isotope monitored: ^{114}Cd*

Water-Soluble Proteins

Higher molecular weight compounds, referred to as protein complexes with Cd or Ni have been extracted with water. The extracts have then been centrifuged and

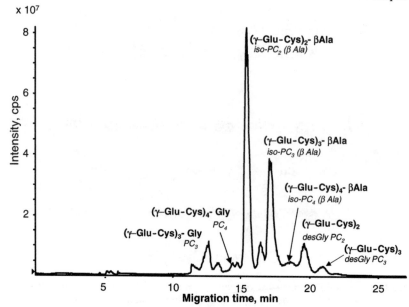

Figure 16.3 *Total ion current CZE-ES MS electropherogram of the Cd containing fraction of a soybean extract. The peaks are labelled with sequences determined by on-line CID MS*

the proteins fractionated by successive filtration through membranes with molecular weight cut-offs of 30 000, 5 000 and 500[20,21] prior to off-line GF AAS. Günther *et al.*[22,23] have used low pressure semi preparative chromatography to separate water-soluble Cd and Zn species. Atomic absorption spectrometry,[20,21,24] total reflection XRF,[22] γ-ray spectrometry[25] and ICP AES or ICP MS[26] have been used to determine the metal concentration in the eluted fractions.

Metalloenzymes and metalloproteases with Pb or Zn as the central metal ions have been investigated.[27] Iron and nickel complexes in bacteria with myoglobin and algal superoxide dismutase have been examined by SDS PAGE with PIXE detection.[28,29] The use of concentrated surfactants and dithiotreitol (DTT) is necessary for the solubilisation of high molecular weight proteins and metalloenzymes.[27]

The distribution (speciation) of some metals (Cu, Zn, Ni) and non-metals (P, S) in soybean flour extracts has been studied.[30–32] Different flat-bed electrophoretic techniques, including gel electrophoresis and isoelectric focusing, have been investigated to study the separation of protein-bound Zn, Ni and Cu in soybean flour extracts. In comparison with SEC, gradient gel electrophoresis affords a better separation of the protein fractions, but can not detect proteins of $M_r < 6500$ Da that can be detected by SEC-AAS.[31]

3 Metal Complexes with Polysaccharides

Plants contain significant concentrations of polysaccharides of which the potentially negatively charged oxygen functions can bind cations electrostatically and even chelate them *via* polyhydroxy groups.[4] In comparison with proteins, however, little is known about the relevance of metal coordination to carbohydrates which are the most abundant (by weight) class of compounds in the biosphere. Attention has been attracted by a structurally complex pectic polysaccharide, rhamnogalacturonan-II (RG-II).[33–36] This ubiquitous component of primary plant cell walls forms dimers cross-linked by 1:2 borate diol esters (dRG-II) that have been found to complex *in vitro* with specific divalent cations[35] and the majority of Pb, Sr, Ba and REEs in fruit and vegetables.[17,37]

Size exclusion chromatography with parallel ICP MS and refractometric detection has been the primary technique used to investigate metal complexes with polysaccharides in plants and related samples.[17,37,38] A typical multielement chromatogram showing the presence of water-soluble metal complexes with dRG-II is shown in Figure 16.4.

Aqueous extracts of plants contain the dRG-II complex but the presence of water-soluble polysaccharide species with higher molecular weights has also been demonstrated.[17] The latter are probably of pectic origin since they can be readily decomposed by enzymic hydrolysis with a mixture of pectinase and hemicellulase to release the dRG-II complex.[17] The same mixture of enzymes has been reported to be efficient in extracting dRG-II-metal complexes from water-insoluble residues of vegetables owing to the destruction of the original pectic structure.[17]

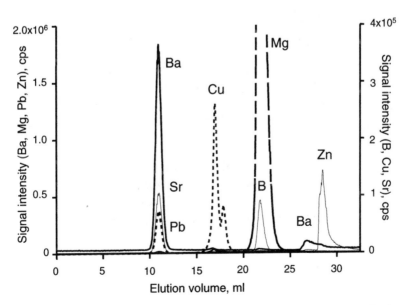

Figure 16.4 *A multielemental SEC-ICP MS chromatogram of a red wine sample showing the presence of RG-II complex with several metals*

4 Metal complexes with phytometallophores

Among the ligands potentially responsible for binding metals in plants an important role is played by hydroxy acids (*e.g.* citric or malic acid) and non-proteinaceous amino acids, referred to as metallophores.[39] Present in root exudates they play a role in solubilising essential elements (*e.g.* Fe(III) from soils[39] but evidence of their action (especially that of nicotianamine) in binding metals in hyperaccumulating plants is appearing.[40] The most widely known mugineic acid family of metallophores is shown in Figure 16.5.[39]

The determination of intact metal-chelate complexes has rarely been successful.[40] Usually phytometallophores were demetallated and analysed by reversed phase HPLC after pre-column derivatisation.[39] For the analysis of intact complexes SEC-ICP MS performs remarkably well offering well resolved chromatograms. A chromatogram showing the elution of a Ni-nicotianamine complex in an aqueous extract of the latex of a hyperaccumulating tree is shown in Figure 16.6a.[40] Note that certain complexes, *e.g.* Ni citrate are not sufficiently stable to elute from a size exclusion column and decompose by sorption of the metal ion on the chromatographic packing.[40]

Iron species in beverages have been investigated (Figure 16.6b).[41] The predominant species in apple juice are Fe(II)-malate and Fe(III)-citrate, whereas Fe(II) and Fe(III) tartrates are the most abundant in white wine.[41]

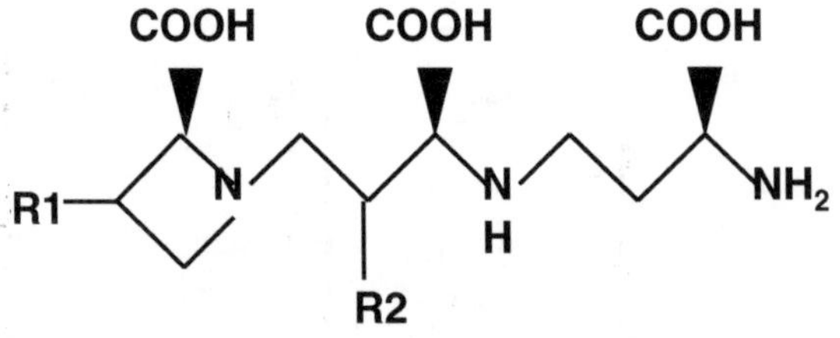

R1 = H   R2 = β-OH   mugineic acid
R1 = H   R2 = H   2'-deoxymugineic acid
R1 = β-OH   R2 = β-OH   3-hydroxymugineic acid
R1 = α-OH   R2 = β-OH   3-*epi*-hydroxymugineic acid

Figure 16.5 *Phytometallophore species in plants*
(from ref. 39)

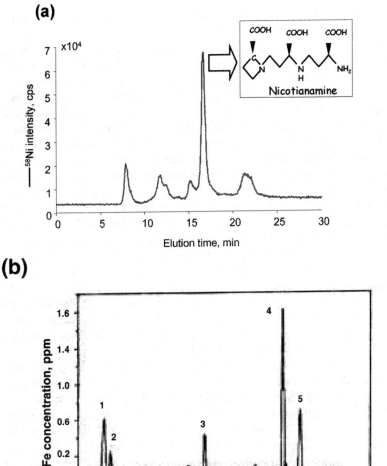

Figure 16.6 *Hyphenated techniques for the determination of metal complexes with organic acids: a) analysis of the latex of a Sebertia tree by size exclusion chromatography* (from Ref. 40); *b) analysis of a plant extract by anion exchange HPLC-AAS: 1,2 – amino acid species of iron, 3 – unknown, 4 – Fe-mugineic acid, 5 – Fe-3-hydroxy-mugineic acid*
(b reproduced with permission from *Fresenius' J. Anal. Chem.*, 2002, **373**, 767, copyright 2002, Springer-Verlag)

5 Other Metal Species in Plant Tissues

The above cases concern reports in which the authors manage to identify the metal-biomolecule complexes eluted from a chromatographic column or at least claim to have done that. There have been, however, a number of studies of the screening of plant extracts for the presence of stable metallospecies without the unambiguous identification of the latter.

Several papers have discussed the speciation of platinum found in plant materials as the consequence of the use of this metal in automotive catalysts.[42-46] Pt has been found to form a number of compounds in the 400–800 Da range but the determination of the molecular weight is highly speculative.[43] The strong correlation of the Pt concentration with the intensity of a pulsed amperometric detection signal (selective for carbohydrates) has enabled the Pt-binding ligands to be identified as carbohydrate oligomers (1000 Da).[42] From the UV absorbance and electrochemical data these species can be characterised as partly oxidised oligosaccharides (about 2–5 monomeric units of aldonic, aldaric and uronic acids) which can originate from the hydrolysis of biopolymers such as pectin (poly-galacturonic acid).[42] A Pt-binding fraction of 180–195 kDa can be isolated from grass. In grass grown on Pt-rich soil seven species ranging from 19 to 1000 kDa can be observed; most Pt is bound to the low molecular-weight species.[45]

Speciation of trace elements, especially aluminium, in tea infusions has attracted some attention.[47,48] The metal is found to be bound to the same range of organic molecules in the infusions, regardless of the origin of the tea.[48] The combination of size exclusion and CE HPLC has allowed the identification of large polyphenolic compounds present in tea as the principal metal-binding organic ligands.[47] Size exclusion LC-ICP MS has also been successfully applied to the fractionation of lead and cadmium complexes in cocoa.[49]

The possible presence of a Cr(III)-β-nicotinamide adenine dinucleotide phosphate (NADP) complex in extracts of Cr-enriched yeast has been investigated.[50] Two Cr samples have been detected by HPLC, one of which has been subsequently identified as the Cr(III)-NADP complex by spiking the yeast extract with a standard.[50] Two low molecular weight anionic Cr species have been detected in cytosolic yeast extracts by ion pair RP HPLC.[51] The coordination bonds of compounds from trivalent cations are prone to hydrolysis and easily deteriorate by ligand exchange reactions.

References

1. M.N.V. Prasad and J. Hagemeyer, *Heavy Metal Stress in Plants – From Molecules to Ecosystem*, Springer, Heidelberg, 1999.
2. R.R. Brooks, M.F. Chambers, L.J. Nicks and B.H. Robinson, *Trends Plant Sci.*, 1998, **3**, 359 .
3. M. Mejare and L. Bulow, *Trends Biotechnol.*, 2001, **19**, 67.
4. D.M. Whitfield, S. Stoijkovski and B. Sarkar, *Coord. Chem. Rev.*, 1993, **122**, 171.
5. J. Szpunar, *Analyst*, 2000, **125**, 963.
6. A. Makarov and J. Szpunar, *Analysis*, 1998, **26**, M44.
7. J. Szpunar, *Trends Anal. Chem.*, 2000, **19**, 127.
8. H. Chassaigne, V. Vacchina and R. Lobinski, *Trends Anal. Chem.*, 2000, **19**, 300.

9. E. Rauser, *Ann. Rev. Biochem.*, 1990, **59**, 61.
10. M.H. Zenk, *Gene*, 1996, **179**, 21.
11. K. Takatera and T. Watanabe, *Anal. Sci.*, 1992, **8**, 469.
12. K. Takatera, N. Osaki, H. Yamaguchi and T. Watanabe, *Anal. Sci.*, 1994, **10**, 567.
13. K. Takatera, N. Osaki, H. Yamaguchi and T. Watanabe, *Anal. Sci.*, 1994, **10**, 907.
14. K. Takatera and T. Watanabe, *Anal. Sci.*, 1993, **9**, 19.
15. I. Leopold and D. Guenther, *Fresenius' J. Anal. Chem.*, 1997, **359**, 364.
16. I. Leopold, D. Guenther and D. Neumann, *Analysis*, 1998, **26**, M28.
17. J. Szpunar, P. Pellerin, A. Makarov, T. Doco, P. Williams and R. Lobinski, *J. Anal. At. Spectrom.*, 1999, **14**, 639.
18. V. Vacchina, K. Polec and J. Szpunar, *J. Anal. At. Spectrom.*, 1999, **14**, 1557.
19. S. Mounicou, V. Vacchina, J. Szpunar, M. Potin-Gautier and R. Lobinski, *Analyst*, 2002, **126**, 624.
20. K. Lange Hesse, *Fresenius' J. Anal. Chem.*, 1994, **350**, 68.
21. K. Lange Hesse, L. Dunemann and G. Schwedt, *Fresenius' J. Anal. Chem.*, 1994, **349**, 460.
22. K. Guenther and A. Von Bohlen, *Spectrochim. Acta*, 1991, **46B**, 1413.
23. K. Guenther, A. von Bohlen and C. Strompen, *Anal. Chim. Acta.*, 1995, **309**, 327.
24. K. Lange Hesse, L. Dunemann and G. Schwedt, *Fresenius' J. Anal. Chem.*, 1991, **339**, 240.
25. A.V. Harms and J.T. van Elteren, *J. Radioanal. Chem.*, 1998, **228**, 139.
26. H. Fingerová and R. Koplik, *Fresenius' J. Anal. Chem.*, 1999, **363**, 545.
27. I. Leopold and B. Fricke, *Anal. Biochem.*, 1997, **252**, 277.
28. Z. Szokefalvi-Nagy, *Nucl. Instrum. Methods Phys. Res., Sect. B.*, 1996, **238**, B110234.
29. Z. Szokefalvi-Nagy, C. Bagyinka, I. Demeter, K. Hollos-Nagy and K.L. Kovacs, *Fresenius' J. Anal. Chem.*, 1999, **363**, 469.
30. J. Schoeppenthau, J. Nolte and L. Dunemann, *Analyst*, **121**, 845.
31. L. Dunemann and H. Reinecke, *Fresenius' J. Anal. Chem.*, 1989, **334**, 743.
32. H. Reinecke and L. Dunemann, *Fresenius' J. Anal. Chem.*, 1990, **338**, 630.
33. T. Ishii and T. Matsunaga, *Carbohydr. Res.*, 1996, **284**, 1.
34. M. Kobayashi, T. Matoh and J.L. Azuma, *Plant Physiol.*, 1996, **110**, 1017.
35. M.A. O'Neill, D. Warrenfeltz, K. Kates, P. Pellerin, T. Doco, A.G. Darvill and P. Albersheim, *J. Biol. Chem.*, 1996, **271**, 22923.
36. P. Pellerin, T. Doco, S. Vidal, P. Williams, J.-M. Brillouet and M.A. O'Neill, *Carbohydr. Res.*, 1996, **290**, 183.
37. J. Szpunar, P. Pellerin, A. Makarov, T. Doco, P. Williams, B. Medina and R. Lobinski, *J. Anal. At. Spectrom.*, 1998, **13**, 749.
38. T. Matsunaga, T. Ishii and H. Watanabe, *Anal. Sci.*, 1996, **12**, 673.
39. M.S. Wheal, L.I. Heller, W.A. Norvell and R.M. Welch, *J. Chromatogr. A*, 2001, **942**, 177.
40. D. Schaumlöffel, L. Ouerdane, B. Bouyssiere and R. Lobinski, *J. Anal. At. Spectrom.* 2003, **18**, 120.

41. G. Weber, *Fresenius' J. Anal. Chem.*, 1991, **340**, 161.
42. F. Alt, J. Messerschmidt and G. Weber, *Anal. Chim. Acta.*, 1998, **359**, 65.
43. N. Jakubowski, C. Thomas, D. Klueppel and D. Stuewer, *Analysis.*, 1998, **26**, M37.
44. D. Klueppel, N. Jakubowski, J. Messerschmidt, D. Stuewer and D. Klockow, *J. Anal. At. Spectrom.*, 1998, **13**, 255.
45. J. Messerschmidt, F. Alt and G. Toelg, *Anal. Chim. Acta.*, 1994, **291**, 161.
46. J. Messerschmidt, F. Alt and G. Toelg, *Electrophoresis*, 1995, **16**, 800.
47. K.L. Oedegard, W. Lund, *J. Anal. At. Spectrom.*, 1997, **12**, 403.
48. A.K. Flaten and W. Lund, *Sci. Tot. Environ.*, 1997, **207**, 21.
49. S. Mounicou, J. Szpunar, R. Lobinski, D. Andrey and C.J. Blake, *J. Anal. At. Spectrom.*, 2002, **17**, 880.
50. M. Beran, R. Stahl and M. Beran, Jr., *Analyst*, 1995, **120**, 979.
51. A. Knöchel and G. Weseloh, *Fresenius J. Anal. Chem.*, 1999, **363**, 533.

Speciation of Metal Complexes with Metallothioneins

1 Introduction

Metallothioneins (MTs) are a group of non-enzymatic low molecular mass (6–7 kDa), cysteine-rich metal-binding proteins, resistant to thermocoagulation and acid precipitation.[1-3] They are considered to intervene in the metabolism, homeostatic control and detoxification of a number of essential (Zn, Cu) and toxic (Cd, Hg, As) trace elements. The most widely studied mammalian MTs have been isolated from kidney, liver and brain samples; MTs isolated from mussels have also been reported. Metallothionein shows genetic polymorphism consisting of a variation in the primary structure by the substitution of between 1 and 15 amino acids.[1-3] The sequence of a metallothionein (rabbit liver MT-1) and the spatial structure of its complex with metals is shown in Figure 17.1.

In terms of speciation analysis the metal-MT system presents several challenges that include the identification of metals involved in the complex, calculation of the stoichiometry of the complex, and the precise identification (unique amino acid sequence) of the ligand present. The requirements for a suitable analytical technique include:

(1) selectivity with regard to the different MT isoforms and sub-isoforms.
(2) selectivity with regard to metals complexed by an individual isoform.
(3) sensitivity to cope with non-induced MT levels in real-life samples.
(4) accurate quantification.

The diversity of MT complexes with metals have made the rabbit liver metallothionein preparation available from Sigma an attractive test sample for the development of novel hyphenated techniques. However, among a large number of papers on metallothionein analysis by hyphenated techniques, those in which the identification and quantification of the MT isoforms in a real-life sample have been achieved are still rare. Size exclusion LC-ICP MS has become a routine

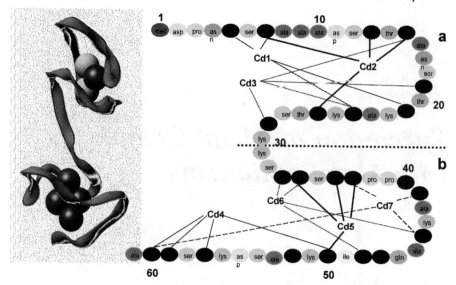

Figure 17.1 *A sequence and the spacial structure of a metallothionein and its complexes with metals; black dots denote cysteine residues*

technique for the screening of raw sample extracts for the MT fraction at concentrations down to the non-induced level. The fraction isolated by preparative SEC then undergoes a more refined characterisation (in terms of the individual isoforms) by reversed phase or anion exchange HPLC-ICP MS or CZE-ICP MS. Electrospray MS is essential for identification of the MT isoforms but the poor sensitivity limits its use to tissues with induced significant MT concentrations. Speciation analysis of MT complexes with metals has been reviewed.[4,5]

This chapter highlights analytical strategies based on hyphenated techniques for speciation of metal complexes with MT in animal and human tissues.

2 Recovery of Metal Complexes with Metallothioneins From Biological Tissues

Soluble extracts of tissues and cultured cells have been prepared by homogenising the sample in an appropriate buffer. Neutral buffers are usually used for the extraction since Zn starts dissociating from protein complexes at pH 5. Cd and Cu are removed at lower pH values. A 10–50 mM Tris-HCl buffer at pH 7.4–9 is the most common choice. For cytosols containing Cd-induced MTs dilution factors up to 10 have been used whereas for those with natural MT levels equal amounts of tissue and buffer are considered to be suitable. The washing of cells with a Tris-HCl buffer (pH 8) containing 1 M EDTA is recommended to remove metal ions reversibly bound to the cell wall.[6]

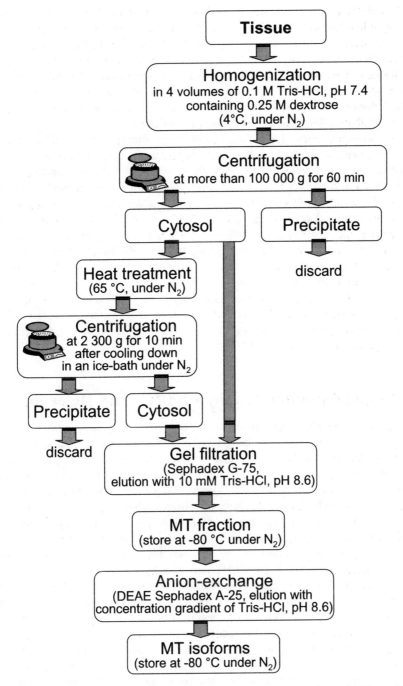

Figure 17.2 *Sample preparation scheme prior to characterisation of metal-MT complexes by hyphenated techniques*
(adapted from Ref. 7)

Metallothioneins are prone to oxidation due to their high cysteine content. Disulphide bridges are then formed and MTs either copolymerise or combine with other proteins to move into the high molecular weight fraction. Therefore the homogenisation of tissues and the subsequent isolation of MTs should be performed in deoxygenated buffers and/or in the presence of a thiolic reducing agent. β-Mercaptoethanol is usually added as the antioxidant. Other components added during homogenisation include 0.02% NaN$_3$ (an antibacterial agent) and phenylmethanesulfonyl fluoride (a protease inhibitor).

The homogenisation step is followed by ultracentrifugation at 4 °C. As a result two fractions, a soluble fraction (cell supernatant, cytosol) and a particulate fraction (cell membranes and organelles) are obtained. The heating of the supernatant at 60 °C for 15 min causes the precipitation (and removal) of the high molecular weight proteins; MTs which are heat stable remain in the supernatant. This procedure decreases the protein load on the HPLC column not only improving the separation of MT isoforms, but also prolonging the column lifetime. Several workers, however, prefer gel filtration to heat treatment for the isolation of the metallothionein fraction from the tissue cytosol. Guidelines for the preparation of biological samples prior to quantification of MTs have been proposed.[7] The typical extraction efficiency from experimental rats and mice tissues is estimated to be 50–80% of Cd.[8] A sample preparation scheme for an animal tissue prior to analysis of the MT-metal complexes by hyphenated techniques is shown in Figure 17.2.[7]

3 Liquid Chromatography with ICP MS Detection

Different separation mechanisms have been employed, each with a different purpose. Size exclusion chromatography tolerates raw sample extracts and is a useful technique to indicate the concentration of MT in the sample and to isolate the MT fraction prior to a more refined characterisation by HPLC using other separation mechanisms or by CZE. The SEC separation efficiency is rather poor, the MT-1 and MT-2 cannot be separated from each other but they both can be separated from the MT-3 fraction.[9] Anion exchange LC offers separation of the MT-1 and MT-2 isoform classes. A more complete separation within each of the classes is preferably carried out by reversed phase HPLC but high resolution chromatograms by AE HPLC-ICP MS have also been reported.[10]

In terms of detection, the induced MT concentrations are sufficiently high to allow a wide range of detection techniques to be used including AAS, ICP AES and ICP MS. Indeed, the metals of concern, Cd, Cu and Zn, are among the most readily detectable by AAS. The added value of ICP AES is the possibility of simultaneous monitoring of sulfur. ICP MS using a quadrupole analyser is becoming a standard detection technique despite the poor detection capability for sulfur.[4,5] The use of ICP TOF MS has been reported[10] but the advantages in comparison with a quadrupole analyser are not evident. The use of a sector-field instrument in the medium resolution mode (3000) allows the simultaneous detection of metals and sulfur at the expense of sensitivity for metals.[11] The low resolution mode of sector field instruments offers detection limits sufficiently low to determine the MT concentrations in non-induced samples. However, the control of contamination and analyte interactions with the column stationary phase at these levels are becoming a challenge.[12]

The tabulated experimental conditions for HPLC with different detectors for speciation analysis of metal complexes with metallothioneins in different animal tissues can be found in the review papers.[4,5]

Size Exclusion LC-ICP MS

The first step of a measurement procedure consists of verification whether a metal (*e.g.* Cd)-binding ligand was bio-synthesized as a response of an organism to metal stress. This can be achieved by the comparison of a size exclusion-ICP MS chromatogram obtained for a tissue sample of a control (unexposed) animal with a corresponding chromatogram obtained for a tissue sample of an exposed animal (Figure 17.3). In a case where the formation of a novel stable metal complex with a bioligand present in the sample is demonstrated, it should be verified whether this ligand was originally present in the sample or whether it was bioinduced as an organism's response to metal stress. For this purpose a SE-ICP MS chromatogram is acquired for an extract of a sample that has been spiked with an

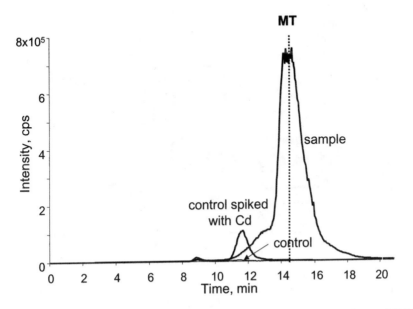

Figure 17.3 *Detection of the bio induction of metallothionein by size exclusion LC-ICP MS*
(from ref. 21)

excess of Cd just after dissection and compared with the chromatogram of an extract of the tissue sample of an exposed animal.

The area of the peak corresponding to the MT fraction is proportional to the quantity of the metal bound by MT and thus, to the quantity of metallothionein induced. Quantification aspects in size exclusion LC have been discussed by Wolf *et al.*[13] who proposed an elegant way for quantification by calibration with a column by-pass injection of elemental standards.

Preparative size exclusion chromatography is commonly used to isolate and clean up the MT fraction from the tissue cytosols prior to separations using other HPLC mechanisms or capillary electrophoresis.

Anion Exchange HPLC-ICP MS

MT isoforms can be separated by AE LC owing to their negative charge in solution. Weak basic anion exchangers with diethylaminoethyl functional groups (DEAE) have been used. Low pressure AEC can separate the MT-1 and MT-2 classes but not sub-isoforms showing only minor differences in the electrical charge.[14-17] Anion exchange HPLC allows a more accurate characterisation of

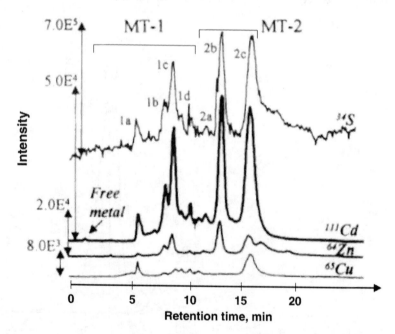

Figure 17.4 *Chromatographic profiles obtained for a rabbit liver metallothionein mixture by anion exchange LC coupled to HR-ICP MS (^{34}S) and Q-ICP MS (metals); letters a, b, c and d correspond to separated isoforms*
(Reproduced from *Anal. Chem.*, 2000, **72**, 5874, copyright 2000, with permission from the American Chemical Society)

MT. It is mostly used with AAS [14–17] or ICP AES detection[13] but ICP MS detection is becoming a standard approach.[10,18] Buffer concentrations up to 250 mM have been reported in the final phase of the gradient[10] but at these concentrations the problem of clogging is unlikely to be avoided in routine analyses. Ogra and Suzuki[18] have proposed a setup in which the MT size exclusion fraction is introduced directly into an anion exchange column through a four-way valve. Figure 17.4 shows a chromatogram obtained by anion exchange HPLC-ICP MS demonstrating the separation of each of the major MT classes, MT-1 and MT-2, into sub-classes.[19]

Reversed Phase HPLC-ICP MS

Reversed phase HPLC with C_4, C_8, or C_{18} columns is the principal separation technique used for the separation of the MT isoforms.[4,20] The MTs are eluted by aqueous buffers with a gradual decrease in polarity produced by the addition of methanol or acetonitrile. The disadvantage of reversed phase separations is the need for an organic modifier requiring the addition of oxygen and the use of Pt-cones in ICP MS. The advantage is the compatibility with ES MS/MS. Elegant separations can be obtained by microbore HPLC with ICP MS detection, the conditions being directly transposable to HPLC-ES MS. A chromatogram of rat tissue extract is shown in Figure 17.5.[21,22]

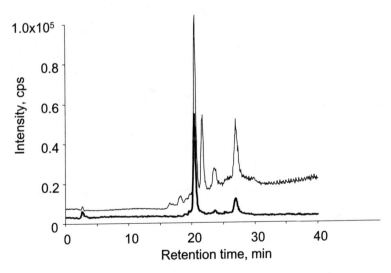

Figure 17.5 *Microbore reversed phase HPLC-ICP MS chromatogram of the rat liver MT fraction. Regular line:* ^{114}Cd. *Bold line* ^{65}Cu

4 Capillary Electrophoresis-ICP MS

Several authors have used metallothionein isolated from rabbit liver[23 27] to optimise the interface design, but both the number of these studies and the quality of the electropherograms reported are incredibly low in comparison with the data available in the literature for CZE separations of metallothionein isoforms using UV detection.[28] A comprehensive summary of these methods in tabular form can be found elsewhere.[5]

The most successful work has been carried out using the CEI-100 interface (*cf.* Chapter 5). MT isoforms have been separated by CE and the elements Cd, Cu, Zn and S have been detected simultaneously using ICP sector field MS in the medium resolution mode. The quantification was achieved using on-line isotope dilution. This was carried out by the continuous introduction of an isotopically enriched, species-unspecific spike solution after the separation step. Based on the known number of S atoms (21) per MT molecule each MT isoform could be quantified independently. Additional measurements of the metals Cu, Zn, and Cd, followed by the determination of the sulfur-to-metal ratio in MT, has enabled the determination of the stoichiometry of a complex. A typical molar flow electro-

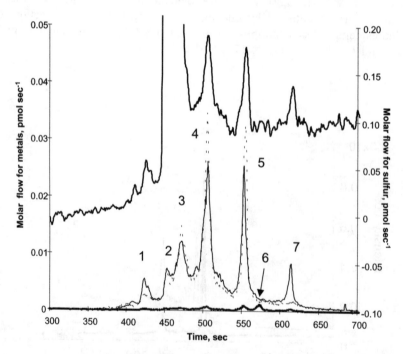

Figure 17.6 *A molar flow electropherogram obtained by CZE-ICP SF MS allowing the determination of the metal-complex stoichiometry using isotope dilution quantification;* ^{114}Cd *(dashed line),* ^{65}Cu *(thin line),* ^{66}Zn *(thick line),* ^{32}S *(regular line) 1,2,3,6-unidentified, 4-Cu_4Cd_3-MT-1, 5-Cu_4Cd_3-MT-2, 7-Cd_7-MT-2*

pherogram allowing a direct reading of the metal stoichiometry is shown in Figure 17.6. The resolution can be further increased by using a coated capillary.[29]

Capillary electrophoresis shows a better separation efficiency of MT isoforms and their complexes in comparison with HPLC. It also suffers from fewer artefacts resulting from metal exchange with the stationary phase or adsorption of MTs which may occur in chromatography. However, the detection limits for sulfur by CZE-ICP HR MS in the medium resolution mode (*ca.* 75 mg L^{-1}) are insufficient for the determination of basal (non-induced) MT levels.[30]

5 Identification of MT Isoforms by Electrospray MS

Neither CZE-ICP MS nor HPLC-ICP MS enables the identification of the MT species detected; HPLC-ES MS needs to be used for this purpose.[22,31] Usually two analyses are run (Figure 17.7).[32] First, ES mass spectra of the eluted complexes are acquired in close to neutral pH and the molecular masses of the metal-MT complexes present are determined. In the second HPLC run the column eluate is acidified on-line and electrospray mass spectra of the demetallated proteins are

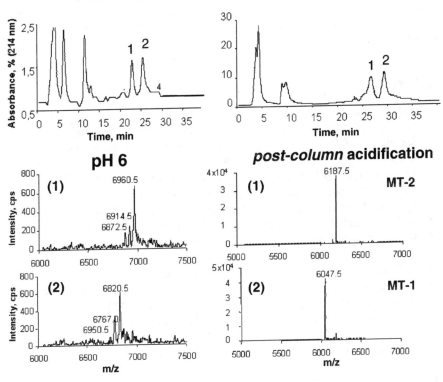

Figure 17.7 *RP HPLC chromatograms of rabbit liver MT isoforms and ES mass spectra corresponding to peaks detected at pH 6 and after post-column acidification (from ref. 32)*

acquired at identical retention times. The demetallation allows a 10-fold decrease in the detection limits so that MT isoforms undetected as metal complexes can be seen.[32]

The separation of demetallated proteins prior to ES MS detection is less popular because of their interactions with the stationary phase. Some successful electrophoretic separations[33,34] have been reported.

6 Analysis of Human and Animal Tissue Samples

As mentioned above, most analytical method development work has been carried out using the relatively well purified MT standards available from Sigma. The maturity of the methodology now enables applications to real-life samples to be reported. These include the characterisation of metal complexes with metallothioneins induced in rat tissues,[21,22] human brain[35] and liver,[13] fish[36] and mussels[10,19,37] used as environmental sentinels.

Applications in the nutrition field need to be mentioned. Enzymic digests of foodstuffs have been analysed in the quest for molecular information to contribute to the knowledge on the bioavailability of essential and toxic elements. Enzymolysis in simulated gastric and gastrointestinal juices has been proposed for meat samples.[38] The soluble fraction of the stomach and upper intestinal contents of guinea pigs on different diets have been investigated for the species of Al, Cu, Zn, Mn, Sr and Rb.

References

1. M.J. Stillman, *Coord. Chem. Rev.*, 1995, **144**, 461.
2. M.J. Stillman, C.F. Shaw and K.T. Suzuki, *Metallothioneins. Synthesis, Structure and Properties of Metallothioneins, Phytochelatins and Metalthiolate Complexes*, VCH, New York, 1992.
3. J.F. Riodan and B.L. Valee, *Metallobiochemistry Part B. Metallothionein and Related Molecules*, Academic Press, New York, 1991.
4. R. Lobinski, H. Chassaigne and J. Szpunar, *Talanta*, 1998, **46**, 271.
5. A. Prange and D. Schaumlöffel, *Anal. Bioanal. Chem.*, 2002, **373**, 441.
6. K. Takatera, N. Osaki, H. Yamaguchi and T. Watanabe, *Anal. Sci.*, 1994, **10**, 907.
7. K.T. Suzuki and M. Sato, *Biomed. Res. Trace. Elem.*, 1995, **6**, 51.
8. H.M. Crews, J.R. Dean, L. Ebdon and R.C. Massey, *Analyst*, 1989, **114**, 895.
9. A. Richarz and P. Brätter, *Anal. Bioanal. Chem.*, 2002, **372**, 412.
10. C.N. Ferrarello, M.M. Montes Bayon, M.R. Fernandez de la Campa and A. Sanz-Medel, *J. Anal. At. Spectrom.*, 2000, **15**, 1558.
11. J. Wang, D. Dreessen, D.R. Wiederin and R.S. Houk, *Anal. Biochem.*, 2001, **288**, 89.
12. C.N. Ferrarello, M.R. Fernandez de la Campa, C. Sariego Muniz and A. Sanz-Medel, *Analyst*, 2000, **125**, 2223 .
13. C. Wolf, U. Rösick and P. Brätter, *Fresenius' J. Anal. Chem.*, 2000, **368**, 839.

14. L.D. Lehman and C.D. Klaassen, *Anal. Biochem.*, 1986, **153**, 305.
15. K.A. High, R. Azani, A.F. Fazekas, Z.A. Chee and J.S. Blais, *Anal. Chem.*, 1992, **64**, 3197.
16. A.H. Pan, F. Tie, B.G. Ru, L.Y. Li and T. Shen, *Biomed. Chromatogr.*, 1992, **6**, 205.
17. P. Sun, X. Shan, Y. Zheng, L. Jin and W. Xu, *J. Chromatogr. Biomed. Appl.*, 1991, **110**, 73.
18. Y. Ogra and K.T. Suzuki, *J. Chromatogr. A*, 1999, **735**, 17.
19. C.N. Ferrarello, M.R. Fernandez de la Campa, J.F. Carrasco and A. Sanz-Medel, *Anal. Chem.*, 2000, **72**, 5874.
20. M.P. Richards, *Meth. Enzym.*, 1991, **205**, 217.
21. K. Polec, M. Peréz-Calvo, O. García-Arribas, J. Szpunar, B. Ribas-Ozonas and R. Lobinski, *J. Anal. At. Spectrom.*, 2000, **15**, 1363.
22. K. Polec, M. Peréz-Calvo, O. Garcia-Arribas, J. Szpunar, B. Ribas-Ozonas and R. Lobinski, *J. Inorg. Biochem.*, 2001, **88**, 197.
23. S.A. Baker and N.J. Miller Ihli, *Appl. Spectrosc.*, 1999, **53**, 471.
24. V. Majidi and N.J. Miller Ihli, *Analyst*, 1998, **123**, 809.
25. K.A. Taylor, B.L. Sharp, D.J. Lewis and H.M. Crews, *J. Anal. At. Spectrom.*, 1998, **13**, 1095.
26. Q.H. Lu and R.M. Barnes, *Microchem. J.*, 1996, **54**, 129.
27. Q. Lu, S.M. Bird and R.M. Barnes, *Anal. Chem.*, 1995, **67**, 2949.
28. J.H. Beattie, *Talanta*, 1998, **46**, 255.
29. Z. Wang and A. Prange, *Anal. Chem.*, 2002, **74**, 626.
30. D. Schaumloeffel, A. Prange, G. Marx, K.G. Heumann and P. Braetter, *Anal. Bioanal. Chem.*, 2002, **372**, 155.
31. H. Chassaigne and R. Lobinski, *Anal. Chem.*, 1998, **70**, 2536.
32. H. Chassaigne and R. Lobinski, *J. Chromatogr. A*, 1998, **829**, 127.
33. X. Guo, H.M. Chan, R. Guevremont and K.W.M. Siu, *Rapid Commun. Mass Spectrom.*, 1999, **13**, 500.
34. J.H. Beattie, A.M. Wood and G.J. Duncan, *Electrophoresis*, 1999, **20**, 1613.
35. A. Prange, D. Schaumlöffel, P. Brätter, A. Richarz and C. Wolf, *Fresenius' J. Anal. Chem.*, 2001, **371**, 764.
36. H. De Smet, B. De Wachter, R. Lobinski and R. Blust, *Aquatic Toxicol.*, 2001, **52**, 269.
37. C.N. Ferrarello, M.R. Fernandez de la Campa, J.F. Carrasco and A. Sanz-Medel, *Spectrochim. Acta*, 2002, **57B**, 439.
38. L.M.W. Owen, H.M. Crews, R.C. Hutton and A. Walsh, *Analyst*, 1992, **117**, 649.

CHAPTER 18

Speciation of Metal Complexes in Human Body Fluids and Tissues

1 Introduction

Speciation affects the bioavailability and toxicity of elements and so is important in toxicology and nutrition. Progress in the understanding of the functions of metals in metalloproteins, enzymes and nucleic acids is determined by the availability of information on metal species in a complex bioligand environment (*e.g.* blood serum). Speciation studies of trace elements associated with proteins are necessary to elucidate the roles that trace elements play in the structures and functions of biological macromolecules.

The greatest interest in the field of biomedical speciation analysis has been attracted by some essential metals, such as Fe, Cu, and Zn, and toxic metals, such as Al, Cr, Pb, Cd and Hg. Their speciation has usually been investigated in blood (subdivided by centrifugation into plasma (serum) and red cells (erythrocytes)) and in breast milk. Liver and kidney have been the most widely studied organs because of their crucial function in the metabolism of toxic metals, such as cadmium. The speciation of arsenic and selenium metabolites in urine is discussed in the chapters devoted to these elements. The same applies to analytical techniques for selenoproteins which are discussed in Chapter 15.

In clinically relevant samples the metals are distributed among one or more macromolecular species with generally a small low molecular mass fraction, including amino acids and thiols, which is difficult to define because of the complex composition of the biological fluid. In the macromolecular fraction the metals can be an integral part of metalloproteins and metalloenzymes or bind less firmly to transport proteins (albumin, transferrin). In general, low molecular mass complexes of metal ions in body fluids are exchangeable among the plethora of available biological ligands, and many metal ions adsorb or bind non-specifically to multiple low affinity sites on proteins. Therefore, the basic problem of clinical speciation analysis is the definition of an analyte species. Nevertheless, a number of metal complexes with biological ligands seem to be sufficiently stable to pass through a chromatographic column and to be analysed by HPLC-ICP MS.

Because of the complexity of the bioligand environment and equilibria present, limits need to be set in this chapter regarding the use of the term speciation. The complex of a metal with a given protein with a unique amino acid sequence and a global tertiary structure should be considered a unique species even though the sample will contain an ensemble of proteins in different states of protonation and local conformation. Hence, speciation describes the distribution of an element among different proteins (*e.g.* the distribution of iron among transferrin, ferritin and heme proteins), or even among isoforms of the same protein. The philosophy of trace element speciation in clinical chemistry and nutritional sciences has been discussed.[1–5]

The scope of this chapter is oriented towards the use of hyphenated techniques. They should not, however, be considered without regard to sampling and sample preservation prior to chromatography. These steps are particularly critical in clinical chemistry because of the low concentrations involved (risk of contamination), the thermodynamic instability of some species, and the complexity of the matrix. Problems that occur during sampling, storage and sample preparation have been extensively discussed.[1,3,6]

2 Analytical Techniques

The most commonly addressed analytical problems include:

(1) screening for the presence of thermodynamically stable metal complexes in a sample.
(2) the high resolution separation of complexes with different proteins.
(3) characterisation of the low molecular weight fraction.

Each of these requires a suitable separation technique. ICP MS is the usual detection technique, although a remarkably large quantity of pertinent information is still acquired by GF AAS analysis of fractions collected from chromatographic columns. Identification, characterisation and determination of metal-binding proteins by liquid chromatography[7] and capillary electrophoresis[8] have been reviewed.

Screening for Stable Metal Complexes with Proteins by Size Exclusion ICP MS

SE LC with ICP MS[9–12] or ICP AES[13,14] detection are the most convenient methods for the fractionation of metal complexing proteins in serum[9,10,13] and in milk.[10–12,14] Multielement detection is feasible but these techniques have also been used for the determination of species of a single element. SEC-ICP MS with a magnetic sector mass spectrometer has been used for the analysis of trace element species in bovine and human serum[15] and in liver extracts.[16]

The sample preparation is usually straightforward. Serum is obtained from whole blood by centrifugation. It is a complex matrix but has the advantage that all the components are soluble in aqueous buffers. Dilution with a chromatographic buffer $(1 + 5)$ and filtration through a 0.2 μm filter are the usual steps prior to injection on the column. Milk is analysed after the elimination of lipids by centrifugation at 3000 rpm for 30 min at 4–5 °C. The resulting whey is injected onto the column; the precipitation of casein with 1 M acetate is optional.[17]

The quality of the chromatograms reported is generally rather poor because of the poorly controlled interactions between the column stationary phase and the species at the picogram level and because of the relatively low resolution of SEC. Another problem is that information on the species identity provided by this type of analysis is limited to a (rough) estimation of the molecular weight of the complexes. A multielement chromatogram obtained for a breast milk sample is shown in Figure 18.1.

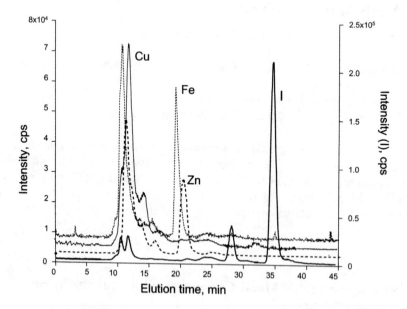

Figure 18.1 *Multielement analysis for metal species in breast milk by size exclusion HPLC-ICP MS*
(Reproduced from *Trends Anal. Chem.*, 2000, **19**, 127, copyright 2000, with permission from Elsevier)

Differentiation Between Albumin and Transferrin-Bound Metals by AE HPLC-ICP MS

In human body fluids, a good deal of speciation analysis concerns the distribution of a metal between albumin or transferrin. Albumin binding is prominent because this protein is present in serum at far higher concentrations (approaching 1 mM) than other proteins. Transferrin is a Fe^{3+} carrier which also binds other ions such as Al^{3+} or Ga^{3+}. Chromium in plasma of healthy people is known to be mainly bound to transferrin and to a lesser degree to albumin.[18,19]

Size exclusion chromatography is incapable of separating albumin from transferrin. Anion exchange HPLC has been proposed where the separation of albumin and transferrin is required.[20–23] The separation of human serum transferrins with different iron-binding states has been reported using a pyridinium polymer column.[24]

Analysis of the Low Molecular Weight Fraction in Serum

When low molecular weight compounds are of interest, sample preparation by ultrafiltration on a 10 kDa filter is required.[25–27,28] The obtained filtrate is protein free and can be analysed, *e.g.* for drug metabolites or porphyrins. Reversed phase chromatography with ICP MS detection is the preferred technique for the metal species in the filtrate.[25–27] Dialysis and purification of serum samples by SE HPLC are required if further separation by RP HPLC is to be undertaken.[17] The presence of aluminium citrate in human serum has been identified by AE chromatography and confirmed by (off-line) ES MS/MS.[29]

3 Overview of Applications

Toxic Elements in Serum

Aluminium build-up in patients with chronic renal failure results in a number of diseases. Efforts have been made to elucidate the mechanisms of this build-up by the investigation of the Al speciation in serum.[20–23,30–35] The techniques used for speciation studies of Al in biological fluids have been reviewed.[31,34] Transferrin seems to bind ~90% of the total serum Al;[1] the remaining 10% is likely to form a complex with citrate.[29] The addition of a desferrioxamine (DFO) drug to serum leads to the displacement of Al from transferrin and to the formation of a Al/DFO chelate.[21] The interaction between Al and Fe for binding to transferrin and the effect of citrate has been studied.[31] As the Al level increases in occupationally exposed individuals, more Al is bound to the high molecular mass fractions.[32]

The use of a HR ICP MS detector has enabled Al speciation to be performed at the basal levels in human serum.[23] The binding of serum proteins to Al and Si has been studied by further separating the HPLC column fractions using SDS-PAGE.[20] The low molecular mass Al complexes were separated by ultrafiltration using a 10 kDa membrane.[32]

The interest in chromium speciation in clinical chemistry results from markedly

204

Chapter 18

elevated plasma Cr concentrations in dialysis patients and workers exposed to elevated Cr levels in the workplace.[18,19] The kinetics of Cr(III) absorption from the dialysis fluid is controlled by the time-dependent binding of the element by albumin and transferrin.[18]

Lead in human serum has been found in at least three molecular weight fractions. The major part of lead is coincident with copper and found to be bound to ceruloplasmin. Of the protein-bound Pb recovered, 80% is reported to be contained in proteins with an apparent molecular mass of 240 kDa, and 20% in proteins with an apparent molecular mass of 45 kDa.[36]

Essential Elements in Serum

Cu, Fe and Zn are essential elements and constituents of important metalloproteins such as ferritin (Fe, Cu, Zn), myoglobin (Fe), cytochrome (Fe), and metalloenzymes: β-amylase (Cu), alcohol dehydrogenase (Cd, Zn) and carbonic anhydrase (Cu, Zn). They have been determined in a multielement array by SE HPLC with ICP MS or ICP AES detection. The binding of Fe to transferrin has been studied by HPLC-GF AAS.[22] Immunoaffinity chromatography compares favourably with SE HPLC for the isolation of proteins binding Zn in human serum.[37] The Zn-containing serum proteins are separated by PAGE; the metal detected by INAA.[38]

In the low molecular fraction Cu, Fe, and Zn exist as chelates with porphyrins. The separation and quantification of these complexes are of biological and clinical significance because abnormal species and/or excessive amounts of porphyrins are associated with a variety of disorders.[39] Methods for the separation and determination of porphyrins, their precursors, their isomers and their free acid and ester derivatives, metalloporphyrins and haematoporphyrin and its derivatives in biological matrices have been reviewed.[39] The majority of methods are based on HPLC with spectrophotometric or fluorescence detection but the use of the HPLC-ICP MS coupling allows the simplification of the sample preparation procedure.[40] The high detection sensitivity makes a single extraction step prior to chromatography sufficient and renders derivatisation unnecessary.[40]

Vanadium is considered as an environmental hazard but it is also regarded as a therapeutic agent in diabetes treatment. Vanadium (V) binding is very weak and very sensitive to changes in its environment. The original pH value of the fluid should be adhered to during chromatography. Reversed phase HPLC with acetonitrile is not recommended because of the risk of denaturation of metal complexes during separation. Very mild separation conditions are necessary to avoid the formation of artefacts.[41] The separation of protein-bound vanadium in incubated serum samples has been discussed.[42]

Speciation of Si in biological fluids has been reviewed.[31,34] Si was not bound specifically to any serum protein[20] but adsorbed *via* weak interactions of silicic acid.[1] Ultrafiltrable Si constitutes 15–45% of this element.

Magnesium species in human serum have been studied by AE HPLC with GF

AAS detection;[43] the metal was found to be associated with both the albumin and globulin fractions but not with transferrin.

Free iodine has been determined in serum fractionated by SE HPLC.[44] A SE HPLC-ICP MS chromatogram showing four iodine species in human serum has been reported.[45]

Trace Elements in Erythrocytes

Erythrocytes (packed cells, red blood cells) need to be lysed to free their content prior to chromatographic separation. Three freeze-thaw cycles to lyse the cells have been proposed followed by a ten-fold dilution with a buffer and centrifugation to remove fragments of membranes.[36,46] An alternative procedure recommended by Cornelis *et al.*[3] is based on mixing one part of packed cells with one part of toluene and 40 parts of ice-cold water, followed by centrifugation and 0.45 μm filtration of the lysate.

The distribution of metals in erythrocytes has been studied by SE HPLC with ICP MS or ICP AES detection. The major lead binding site in erythrocyte was identified as haemoglobin.[46] However the latter which represents up to 94% of the total amount of proteins in packed cells complicates the identification of other proteins.[36]

Trace Elements in Breast Milk

Milk is a single source of nutrients for infants and hence is relevant to their nutritional status. Since speciation affects the bioavailability of trace elements, the latter should be present in milk not only in appropriate quantities but also in specific forms. The interest in speciation has been stimulated by the need to match the elemental species present in breast milk to those in infant formulae. Speciation of trace elements in human milk has been the subject of fairly comprehensive multielement studies.[14,47] A number of other studies have concerned specific elements such as Cd,[48] Zn,[28,49] Fe[17] and I.[11] Comparisons between the elemental speciation in human and cow's milk have been made.[11,28]

Casein (a mixture of α, β and k-casein) is the major milk protein. It is present in milk in the micellar form (spherical polyspread colloidal aggregates with molecular weights exceeding 1000 kDa) and elutes in the void.[47] Lactoferrin exists in four molecular forms in nature, and is dominant in human milk.[17] It usually elutes together with casein and immunoglobulins.[47] Another metal-complexing protein is albumin but the signal identification is tricky because of the possible presence of disaggregated micelles of casein.[47] Free amino acids, small complexes such as citrate, and ions such as iodide elute in or close to the total volume of the column. The use of enzymes has been proposed to destroy the protein complexes excluded from the column into smaller proteins that would elute in the middle of the chromatogram; however, the preliminary investigations have not allowed, the identification of the product species.[12]

Trace Elements in Amniotic Fluid

Amniotic fluid is a urine-like fluid inhaled and swallowed by the human fetus. Some heavy metals, *e.g.* Pb, can cross the placenta and end up in amniotic fluid. Metal-binding ligands are important in amniotic fluid because of the potential for being transporters to the neurological system.[50] Prior to analysis an amniotic fluid sample is usually centrifuged and the supernatant stored frozen at -20 °C. Most Pb is found to be bound to ceruloplasmin or to a 5 kDa Zn peptide.[50] The Cu-ceruloplasmin peak is used as an elution volume marker.[50] The poor resolution of SEC prevents the accurate assignment of elements bound to either albumin or transferrin,[50] the two important bioligands in amniotic fluid.[50]

References

1. A. Sanz-Medel, *Spectrochim. Acta,* 1998, **53B**, 197.
2. D.M. Templeton, *Fresenius' J. Anal. Chem.*, 1999, **363**, 505.
3. R. Cornelis, J. De Kimpe and X. Zhang, *Spectrochim. Acta,* 1998, **53B**, 187.
4. R. Cornelis and J. De Kimpe, *J. Anal. At. Spectrom.*, 1994, **9**, 945.
5. S.J. Fairweather-Tait, *Fresenius' J. Anal. Chem.*, 1999, **363**, 536.
6. D. Behne, *Analyst*, 1992, **117**, 555.
7. M.B. de la Calle Guntinas, G. Bordin and A.R. Rodriguez, *Anal. Bioanal. Chem.*, 2002, **374**, 369.
8. M.P. Richards and J.H. Beattie, *J. Cap. Elec.*, 1994, **3**, 196.
9. S.C.K. Shum and R.S. Houk, *Anal. Chem.*, 1993, **65**, 2972.
10. A. Raab and P. Braetter, *J. Chromatogr. B: Biomed Appl.*, 1998, **707**, 17.
11. L. Fernandez Sanchez and J. Szpunar, *J. Anal. At. Spectrom.*, 1999, **14**, 1697.
12. J. Szpunar, *Trends Anal. Chem.*, 2000, **19**, 127.
13. K. Pomazal, C. Prohaska, I. Steffan, G. Reich and J.F.K. Huber, *Analyst*, 1999, **124**, 657.
14. V.E. Negretti de Braetter, S. Recknagel and D. Gawlik, *Fresenius' J. Anal. Chem.*, 1995, **353**, 137.
15. J. Wang, R.S. Houk, D. Dreessen and D.R. Wiederin, *J. Biol. Inorg. Chem.*, 1999, **4**, 546.
16. J. Wang, D. Dreessen, D.R. Wiederin and R.S. Houk, *Anal. Biochem.*, 2001, **288**, 89.
17. Y. Makino and S. Nishimura, *J. Chromatogr. B; Biomed. Appl.*, 1992, **117**, 346.
18. F. Borguet, R. Cornelis and N. Lameire, *Biological Trace Element Res.,*, 1990, 449.
19. R. Cornelis, F. Borguet, S. Dyg and B. Griepink, *Mikrochim. Acta*, 1992, **109**, 145.
20. K. Wrobel, E. Blanco Gonzalez, K. Wrobel and A. Sanz-Medel, *Analyst*, 1995, **120**, 809.
21. A.B. Soldado Cabezuelo, E. Blanco Gonzalez and A. Sanz-Medel, *Analyst*, 1997, **122**, 573.

22. G.F. Van Landeghem, P.C. D'Haese, L.V. Lamberts and M.E. De Broe, *Anal. Chem.*, 1994, **66**, 216.
23. A.B. Solado Cabezuelo, M. Montes Bayon, E. Blanco Gonzalez, J.I. Garcia Alonso and A. Sanz-Medel, *Analyst*, 1998, **123**, 865.
24. K. Harad, A. Kuniyasu, H. Nakayama, M. Nakayama, T. Matsunaga, Y. Uji, H. Sugiuchi and H. Okabe, *J. Chromatogr. B*, 2002, **767**, 45.
25. W.A.J. De Waal, F.J.M.J. Maessen and J.C. Kraak, *J. Chromatogr. A*, 1987, **407**, 253.
26. S.G. Matz, R.C. Elder and K. Tepperman, *J. Anal. At. Spectrom.*, 1989, **4**, 767.
27. G.K. Poon, F.I. Raynaud, P. Mistry, D.E. Odell, L.R. Kelland, K.R. Harrap, C.F.J. Barnard and B.A. Murrer, *J. Chromatogr. A*, 1995, **712**, 61.
28. A. Theobald and L. Dunemann, *J. High Res. Chromatogr.*, 1996, **19**, 608.
29. T. Bantan, R. Milacic, B. Mitrovic and B. Pihlar, *J. Anal. At. Spectrom.*, 1999, **14**, 1743.
30. E. Blanco Gonzalez, A.B. Soldado Cabezuelo and A. Sanz-Medel, *Biomed. Chromatogr.*, 1998, **12**, 143.
31. P.C. d'Haese, G.F. Van Landeghem, L.V. Lamberts and M.E. De Broe, *Mikrochim. Acta*, 1995, **120**, 83.
32. H.B. Roellin and C. Nogueira, *Eur. J. Clin. Chem. Clin. Biochem.*, 1997, **35**, 215.
33. W. Toda, J. Lux and J.C. Van Loon, *Anal. Lett.*, 1980, **13**, 1105.
34. G.F. Van Landeghem, M.E. De Broe and P.C. D'Haese, *Clin. Biochem.*, 1998, **31**, 385.
35. K. Wrobel, E. Blanco Gonzalez and A. Sanz-Medel, *J. Anal. At. Spectrom.*, 1994, **9**, 281.
36. I.A. Bergdahl, A. Schuetz and A. Grubb, *J. Anal. At. Spectrom.*, 1996, **11**, 735.
37. U. Gless, Y. Schmitt and J.D. Kruse Jarres, *Fresenius' J. Anal. Chem.*, 1992, **343**, 88.
38. S.F. Stone, D. Hancock and R. Zeisler, *J. Radioanal. Nucl. Chem.*, 1987, **112**, 95.
39. J.W. Ho, *J. Liq. Chromatogr.*, 1990, **13**, 3741.
40. U. Kumar, J.G. Dorsey, J.A. Caruso and E.H. Evans, *J. Chromatogr. Sci.*, 1994, **32**, 282.
41. K. De Cremer, J. De Kimpe and R. Cornelis, *Fresenius' J. Anal. Chem.*, 1998, **707**, 17.
42. S. Lustig, D. Lampaert, K. De Cremer, J. De Kimpe, R. Cornelis and P. Schramel, *J. Anal. At. Spectrom.*, 1999, **14**, 1357.
43. B. Godlewska-Zylkiewicz, B. Lesniewska and A. Hulanicki, *Anal. Chim. Acta*, 1998, **358**, 185.
44. B. Michalke, P. Schramel and S. Hasse, *Mikrochim. Acta*, 1996, **122**, 67.
45. A. Makarov and J. Szpunar, *Analysis*, 1998, **26**, M44.
46. B. Gercken and R.M. Barnes, *Anal. Chem.*, 1991, **63**, 283.
47. E. Coni, A. Alimonti, A. Bocca, F. La Torre, E. Menghetti, E. Miraglia and S. Caroli, *Trace Elem. Electrolyt.*, 1996, **13**, 26.

48. B. Michalke and P. Schramel, *J. Trace Elem. Electrolytes Health Dis.*, 1990, **4**, 163.
49. B. Michalke, D.C. Muench and P. Schramel, *Fresenius' J. Anal. Chem.*, 1992, **344**, 306.
50. G.S. Hall, E.G. Zhu and E.G. Martin, *Anal. Commun.*, 1999, **26**, 93.

Metal Speciation in Pharmacology: Metallodrugs

1 Introduction

Platinum (cisplatin, carboplatin), ruthenium (fac-RuCl$_3$(NH$_3$)$_3$) and gold (auranofin) compounds are well-known in cancer therapy whereas some other gold compounds (aurithiomalate, aurothioglucose) are important antiarthritic drugs.[1,2] A wide range of Tc compounds (*e.g.* Tc-labelled antibodies, Tc-mercaptoacetyl glycine complex) are used for diagnostic imaging of renal, cardiac and cerebral functions and of various forms of cancer.[3] Gadolinium (III) polyaminopolycarboxylic crown complexes are employed as magnetic resonance imaging contrast reagents.[4] Vanadium and tungsten complexes have been considered for use as insulin mimetics.[5] Chemical structures of some of the metallodrugs investigated are shown in Figure 19.1.

The most important reaction pathways of metallodrugs are assumed to include hydrolysis, and binding to nucleotides, DNA fragments and sulfur-containing biomolecules. Metabolism studies can be carried out *in vitro*, by hydrolysing the drug under physiological conditions or making it react with model molecules, or *in vivo*, by examining the speciation of the drug in the blood of patients that have been administered a therapeutic dose. The related analytical challenges will therefore concern the identification of the products of metallodrug metabolism as well as of their complexes with various biomolecules.

Since the discovery of anticancer properties of metallodrugs, the development of analytical techniques has been of special concern for pharmaceutical and biomedical analysis.[6] The choice of a hyphenated technique strongly depends on the problem to be solved. The studies of metallodrugs fall into four major categories:

(1) studies of the drug purity and stability and chemical transformations that occur under physiological conditions.[7–11]
(2) studies of interactions of drugs and their metabolites with biologically

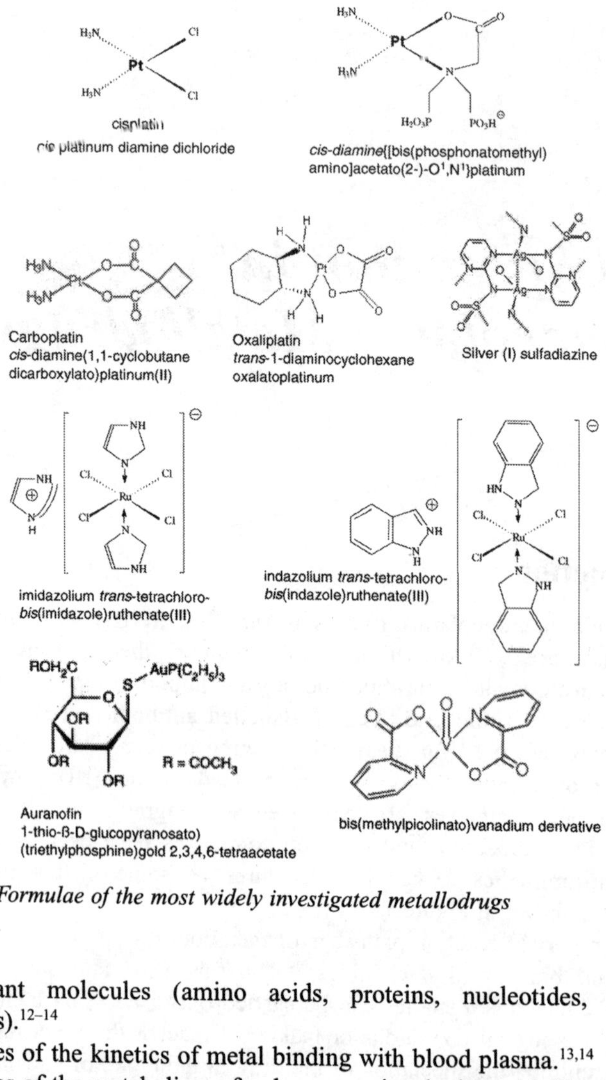

Figure 19.1 *Formulae of the most widely investigated metallodrugs*

relevant molecules (amino acids, proteins, nucleotides, DNA fragments).[12-14]

(3) studies of the kinetics of metal binding with blood plasma.[13,14]

(4) studies of the metabolism of a drug upon incubation with blood plasma.[15]

The analytical chemistry of metallodrugs has been reviewed with particular attention to the determination of Pt species.[6]

2 Analytical Techniques

An appropriate separation technique and sample preparation method need to be chosen as a function of the type of study being undertaken. Consequently, two basic types of analytical approaches have been developed: one for the high molecular weight complexes of metallodrugs with biomolecules based on SE HPLC-ICP MS[10,13] or PAGE-ICP MS;[14] and one for the low molecular weight compounds based on the ultrafiltration of serum followed by analysis of the filtrate by RP HPLC with ICP MS[7,8,10,11] or ES MS[9,16] detection. HPLC conditions for the separation of Pt anticancer drugs, their transformation and reaction products have been tabulated elsewhere.[6]

Stability and Chemical Transformations of Metallodrugs in Model Solutions

Metallodrugs in aqueous solutions are prone to hydrolysis which sometimes leads to intermediate products, *e.g.* the monohydrated complex of cisplatin, $[PtCl(NH_3)_2H_2O)]^+$, which are supposed to be important cytotoxic species. Hence, there is a need to study the evolution of native metallodrugs in physiological solutions and the identification of the products of their hydrolysis.

In saline aqueous solutions metallodrugs exist as a mixture of positively charged and neutral complexes. Because of the chemical properties of the metabolites, reversed phase or cation exchange chromatography have been used.[6] For the separation of these species HPLC with gradients of organic modifier concentration and/or ionic strength can be employed. LC analysis of cisplatin has frequently been performed on strong anion exchange columns or reversed phase columns with a mobile phase containing a lipophilic quaternary ammonium compound.[8] Isolation of the mono- and dihydrated complexes (which are cationic) has been carried out at low pH. The separation of uncharged hydrated complexes on porous graphitic carbon using aq. NaOH as a mobile phase has been reported.[17]

Analytical characterisation is usually not limited by the detection limits and is carried out in matrix free solutions. Therefore electrospray is the detection technique of choice since it allows the identification of the metabolites. On the level of separations, promising results have been reported for separating various Pt drugs and their hydrolytic products by CZE.[18] The same technique is also suitable for the separation of diastereoisomers of lobaplatin.[19] The half-life times for the hydrolytic decomposition of *trans*-$[RuCl_2(im)]$ and *trans*-$[RuCl_4(ind)_2]$ can be determined.[20] No reports of the couplings of CZE with ICP MS or ES MS exist to date but their appearance is only a question of time. The potential of CZE for metallodrug studies has been discussed.[21]

In vitro Studies of Interactions of Metallodrugs with Biologically Relevant Molecules

Thiol amino acids, such as L-methionine or L-cysteine and thiol peptides, such as glutathione in biological fluids, react with significant proportions of administered metallodrugs. The detection and identification of the complexes formed are important in the search for a protective agent against drug-induced toxicity. HPLC-ICP MS serves for the detection of the complexes formed whereas LC-MS is mandatory for their identification.[16]

The hydrated species bind to the guanisine moieties of DNA by means of intra- and interstrand cross-links. The separation and quantification of major Pt-DNA adducts formed by cisplatin using reversed phase HPLC and ^{32}P-postlabelling has been reported.[22] Metal complexes with DNA have been extensively studied by HPLC and CZE. Isomers of platinated oligonucleotides which differ by the binding site of the Pt complex on a single GG sequence have been separated.[12]

Binding of Metallodrugs to Proteins Upon Incubation with Blood Plasma

Figure 19.2 shows schematically the different fates of a metallodrug of interest upon incubation with blood plasma. In addition to being covalently bound to plasma proteins, the drugs undergo a degradation resulting in the formation of a number of metabolites which need to be identified and determined. A point of interest is the determination of the pharmacologically active drug, defined as the intact drug, which is either free or non-covalently bound to plasma proteins.

Earlier studies of the determination of complexes of metallodrugs with proteins in serum have been based on ultrafiltration techniques employing cut-off filters which enable the fractionation of serum into fractions differing in terms of molecular mass, followed by the determination of the metal concentration in the separated fraction off-line by GF AAS or ICP AES.[23] The limited selectivity in terms of the molar mass of the separated complexes can be overcome by the use of size exclusion chromatography which can be directly coupled with ICP MS.[13]

The use of short columns and a careful choice of packing to ensure complete recovery are recommended in order to obtain a rapid evaluation of the protein bound and non-bound drug fractions at a given time. Figure 19.3 shows a chromatogram from a kinetic study of metallodrug binding to blood plasma consituents. The use of longer columns is reported to allow a higher resolution, in terms of the different proteins as well as as the drug metabolites but the losses of the drug or its metabolites on the column may be considerable.[13]

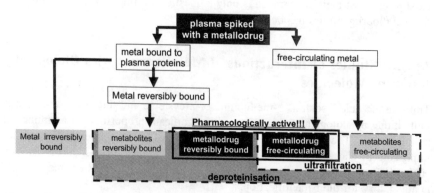

Figure 19.2 *Possible fates of a metallodrug in blood plasma upon administration*

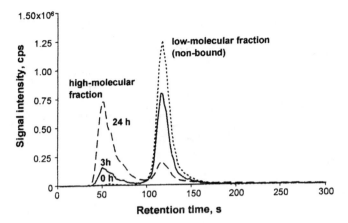

Figure 19.3 *SEC-ICP MS chromatograms (column: Progel TSK column (40 × 4 mm × 6 μm), mobile phase: 30 mM Tris-HCl buffer, pH 7.2) of a serum sample incubated with* cis-*platin*
(reprinted with permission from *Anal. Chim. Acta*, 1999, **387**, 135, copyright 1999, Elsevier)

Protein precipitation has been employed to isolate a fraction of the intact drug together with its metabolites which are reversibly bound to serum proteins. Usually, dilute $HClO_4$ is used for this purpose.[15]

The definition of the total metal found in the ultrafiltrate or in the supernatant after protein precipitation as corresponding to the pharmacologically active fraction of the drug is misleading because several metabolites can be present. Hence an accurate characterisation of the unbound fraction by a hyphenated technique is required.

Metabolism of Metallodrugs Upon Incubation with Blood Plasma

The basic problem in *in vivo* studies is to obtain detection limits that are sufficiently low to match the concentration levels of the unchanged drug several hours after administration. HPLC is the principal separation technique used to achieve the discrimination between the intact drug and its metabolites in the plasma ultrafiltrate or in the supernatant obtained after protein precipitation. ICP MS is the only chromatographic detector offering the sensitivity required. The use of a heated cyclone spray chamber followed by a two stage desolvation allows detection limits down to $0.6\ ng\,mL^{-1}$ to be obtained in reversed phase HPLC-ICP MS.[7] When no desolvation is applied detection limits are $1–2\ ng\,mL^{-1}$.[24] A further decrease in the detection limits can be obtained by eliminating the organic modifier from the mobile phase, reduction of dead volume in the spray chamber and preconcentration by lyophilisation of the supernatant. A chromatogram of a plasma ultrafiltrate allowing the determination of the intact drug (BBR 3364) in the presence of its metabolites is shown in Figure 19.4.

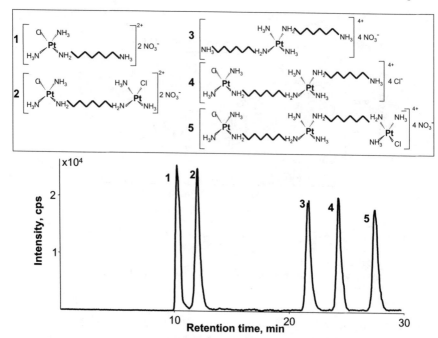

Figure 19.4 *Separation of the platinum-containing drug (5-BBR3464) and its metabolites in the plasma supernatant after deproteinisation by cation exchange HPLC-ICP MS (column: Supelcosil LC-SCX, mobile phase: from 20 to 200 mM of pyridine (pH = 3) in 30 min, isotope monitored: ^{195}Pt)* (from ref. 15)

References

1. B.K. Keppler, *Metal Complexes in Cancer Chemotherapy*, VCH, Weinheim, 1993.
2. E. Wong and C.M. Giandomenico, *Chem. Rev.*, 1999, **99**, 2451.
3. O.K. Hjelstuen, *Analyst*, 1995, **120**, 863.
4. A. Mazzucotelli, V. Bavastello, E. Magi, P. Rivaro and C. Tomba, *Anal. Proc.*, 1995, **32**, 165.
5. J.L. Domingo, *Biol. Trace Elem. Res.*, 2002, **88**, 97.
6. R.R. Barefoot, *J. Chromatogr. B*, 2001, **751**, 205.
7. W.R.L. Cairns, L. Ebdon and S.J. Hill, *Fresenius' J. Anal. Chem.*, 1996, **355**, 202.
8. W.A.J. De Waal, F.J.M.J. Maessen and J.C. Kraak, *J. Chromatogr. A*, 1987, **407**, 253.
9. G.K. Poon, F.I. Raynaud, P. Mistry, D.E. Odell, L.R. Kelland, K.R. Harrap, C.F.J. Barnard and B.A. Murrer, *J. Chromatogr. A.*, 1995, **712**, 61.
10. S.G. Matz, R.C. Elder and K. Tepperman, *J. Anal. At. Spectrom.*, 1989, **4**, 767.

11. Z. Zhao, W.B. Jones, K. Tepperman, J.G. Dorsey and R.C. Elder, *J. Pharm. Biomed. Anal.*, 1992, **10**, 279.
12. H. Troujman and J.C. Chottard, *Anal. Biochem.*, 1997, **252**, 177.
13. J. Szpunar, A. Makarov, T. Pieper, B.K. Keppler and R. Lobinski, *Anal.Chim. Acta*, 1999, **387**, 135.
14. S. Lustig, J. De Kimpe, R. Cornelis and P. Schramel, *Fresenius' J. Anal. Chem.*, 1999, **363**, 484.
15. V. Vacchina, L. Torfi, C. Allievi, R. Lobinski, *J. Anal. At. Spectrom.*, 2003, **18**, 884.
16. O. Heudi, A. Cailleux and P. Allain, *J. Inorg. Biochem.*, 1998, **71**, 61.
17. H.C. Ehrsson, I.B. Wallin, A. Anderson, P.O. Edlund, *Anal. Chem.*, 1995, **67**, 3608.
18. B.W. Wenclawiak and M. Wollmann, *J. Chromatogr. A*, 1996, **724**, 317.
19. C. Vogt and G. Werner, *J. Chromatogr. A*, 1994, **686**, 325.
20. A. Kung, T. Pieper and B.K. Keppler, *J. Chromatogr. B*, 2001, **759**, 81.
21. A.R. Timerbaev, A. Küng, B.K. Keppler, *J. Chromatogr. A*, 2002, **945**, 25.
22. D. Pluim, M. Maliepaard, R.C. van Waardenburg, J.H. Beijnen and J.H. Schellens, *Anal. Biochem.*, 1999, **275**, 30.
23. R.J. Einhäuser, M. Galanski and B.K. Keppler, *J. Anal. At. Spectrom.*, 1996, **11**, 747.
24. P.J. Galletis, J.L. Carr, J.W. Paxton and M.J. McKeage, *J. Anal. At. Spectrom.*, 1999, **14**, 953.

Subject Index